HYPOTHESIS, PREDICTION, AND IMPLICATION IN BIOLOGY

D0503378

JEFFREY J. W. BAKER
George Washington University

GARLAND E. ALLEN
Washington University

HYPOTHESIS, PREDICTION, AND IMPLICATION IN BIOLOGY

ADDISON-WESLEY PUBLISHING COMPANY
Reading, Massachusetts
Menlo Park, California · London · Amsterdam · Don Mills, Ontario · Sydney

This book is in the

Addison-Wesley Series in the Life Sciences

Johns W. Hopkins III, *Consulting Editor*

ISBN 0-201-00482-8
FGHIJKLMN-AL-79876

PREFACE

The gratifying acceptance of and response to our text *The Study of Biology* (Addison-Wesley, 1967) has led to the development of this book. *Hypothesis, Prediction, and Implication in Biology* is intended to be used as a supplementary text in conjunction with either *The Study of Biology* or another basic text.

Part I contains, unchanged, Chapters 2–5 of *The Study of Biology*. These chapters (especially Chapter 3) "set the stage," so to speak, for the approach used throughout the remainder of *The Study of Biology*. They are included because instructors using this book as a supplement to some other text might wish to assign them in place of the more traditional treatments of "the scientific method" often encountered.

In Part II we present, first, an additional example of logical analysis of a scientific paper, to supplement the example given in Chapter 3. Then we present a series of articles and letters which demonstrate the difficulties which may occur when the fruits of scientific research encroach (or show signs of doing so) upon areas which may have sensitive socio-political ramifications.* This material should be of special interest to those students majoring in the social sciences or the humanities, but it is of course, important for the science major as well.

---

* The commission on Undergraduate Education in the Biological Sciences has recently published a book, *Biology for the Non-Major* (available free of charge from CUEBS, Suite 403, 1717 Massachusetts Avenue, N.W., Washington, D.C. 20036). According to this publication, a growing number of biologists feel that students need more exposure to areas in which the results of scientific research influence social and political behavior.

In summary, this book illustrates the functioning of biology both as an investigative science in its own right and as a discipline whose hypotheses may possess far-reaching philosophical and sociological significance. We believe, therefore, that it will form a valuable addition to the reading material of any introductory biology course.

Washington, D.C.   J. J. W. B.
April 1968   G. E. A.

CONTENTS

PART I

CHAPTER 2

MAJOR GENERALIZATIONS IN BIOLOGY

2-1 INTRODUCTION

Generalizations are statements which relate a number of specific and isolated items of information. The statement "All cats have four legs" is a generalization because it says something about a whole category of objects called cats. It relates a specific item of information such as "That white cat has four legs" to another specific item of information, "This gray cat has four legs." Generalizations are very important to everyday thinking. They are no less so to science. Generalizations provide a means of telescoping and condensing experience. They also allow a person to go beyond the specific individual event (for example, that a given cat has four legs) and predict aspects of experience with which he has not had direct contact. In summary, then, generalizations have two important functions: (1) They relate aspects of experience and so provide a way by which people organize their thinking, and (2) they provide a basis for prediction.

2-2 THE CELL CONCEPT

The cell is the basic unit of life. The importance of viewing organisms on the cellular level is perhaps best summarized in the words of a German physiologist of the nineteenth century, Max Verworn:

It is to the cell that the study of every bodily function sooner or later drives us. In the muscle cell lies the problem of the heartbeat and that of muscular contraction;

in the gland cell reside the causes of secretion; in the lining cells of the digestive tract lies the problem of the absorption of food; and the secrets of the mind are hidden in the ganglion [i.e., nerve] *cell.*

What is a cell? We find, first, that cells take many forms. There are nerve cells, muscle cells, digestive cells, skin cells, bone cells, visual cells in the eye, and many others. But though they differ in a variety of ways, most cells have certain fundamental similarities. A simplified diagram of a cell, showing just a few of the essential parts, can be seen in Fig. 2–1. This diagram shows that cells possess an outer limiting membrane, the **cell** or **plasma membrane.** Inside the plasma membrane are the **nucleus** and the **cytoplasm.** The cytoplasm is a watery medium in which a variety of molecules and larger structures called **organelles** (*-elle* = little, thus "little organ") are suspended. The nucleus, which is surrounded by its own limiting membrane, contains nucleic acid, believed to be responsible for the transmission of genetic information.

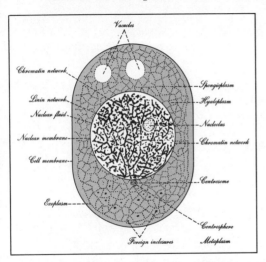

FIG. 2–1. Biology textbooks of the 1920's contained diagrams of "typical" cells like the one at the left. Such drawings were based upon cell anatomy as revealed by the light microscope.

The "cell concept" as viewed today can be summarized in the following four propositions.

1) *Virtually all living organisms are composed of cells.* This generalization extends from simple one-celled organisms such as an amoeba or a bacterium to complex many-celled forms such as a man or a tree. Cells are thus the basic structural units of which organisms are composed.

2) *Cells are the site of all metabolic reactions in an organism.* Even in a complex form such as man, all metabolism takes place within individual cells. Thus, for example, the chemical processes which provide the energy for contraction of a muscle cell take place within the muscle cell itself.

3) *Cells arise only from preexisting cells.* No spontaneous generation of cells occurs. A multicelled organism grows by duplication of its individual cells. By

special cell divisions, some organisms produce gametes, which are capable of generating a whole new organism. The idea that all cells originate only from pre-existing cells is fundamental to the modern cell theory.

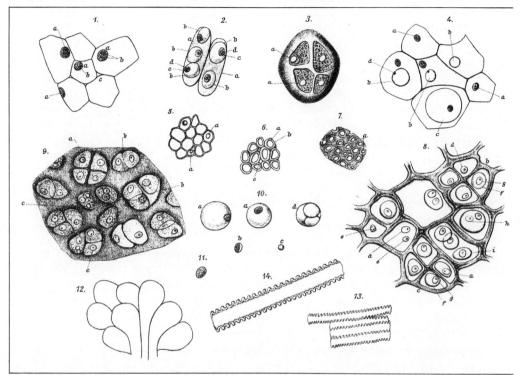

FIG. 2–2. Drawings from Schwann's *Microscopical Investigations* (1839). This work shows that by the mid-nineteenth century a great deal of detail was known about the structure of cells.

4) *Cells contain hereditary material* **(nucleic acid)** through which specific characteristics are passed on from parent cell to daughter cell. The hereditary material contains a "code" which ensures that there will be continuity of species from one cell generation to the next.

The modern cell concept (sometimes called the cell theory) is not the work of any one man. Cells were first observed and described by some of the early seventeenth-century microscopists. Robert Hooke's *Micrographia* (1665) contains some of the first clear plant-cell drawings, made from observations of thin sections of cork.* Hooke coined the term "cell" to refer to these boxlike structures. The fact that all living organisms are composed of cells was not recognized,

---

* The term "cork" refers to the bark, or outer covering layer, of any woody plant. The cork used for thermos bottles is the bark of the cork oak tree, found mostly in Spain.

however, until the nineteenth century. The most important generalized statement about the cellular nature of living organisms was made by two German
biologists, Matthias Schleiden and Theodor Schwann, in 1838 and 1839.
Schleiden, a botanist, and Schwann, a zoologist, studied many types of tissue in
their respective fields. Both came to the conclusion that the cell was the basic
structural unit of all living things. Some of Schwann's diagrams of various animal
cells can be seen in Fig. 2–2. Schleiden and Schwann thus contributed the first
of the propositions listed above. The second and third propositions were added
by the German pathologist and statesman Rudolf Virchow (1821–1902). In his
work *Cellular Pathology* (1858), Virchow spoke of the cell as the basic metabolic
as well as structural unit. In this same work he emphasized the continuity in
living organisms by the statement: "omnis cellula e cellula"—"all cells come
from (preexisting) cells." The last two propositions in the modern cell concept
are more recent additions.

2-3 THE GENE CONCEPT

There are two basic ideas as to the way in which characteristics are determined
in the offspring of any two parent organisms. One view is that the offspring
shows a mixture, or blending, of all the characteristics from each parent. This
theory was believed by many people in the nineteenth century. It holds that such
characteristics as hair color, bone size, or flower color are a blend between the
parental characteristics. In morning glories, for example, a cross between a plant
with red flowers and one with white flowers would be expected to produce a first
generation of plants with pink flowers. Observation shows that this is, indeed,
the case. This idea of **blending inheritance** is analogous to that of mixing two
different-colored paints in one bucket. The resulting organism shows characteristics which are a "blend" between the two parental characteristics.

The theory of blending inheritance explained very well the hereditary process
in many organisms such as the morning glory. But there were some situations
which it did not explain. For example, if several pink (first generation) morning
glories were crossed, the second generation showed red, pink, and white morning
glories in a ratio of 1:2:1. The theory of blending inheritance was unable to
account for the reappearance of both the red and white characteristic. Here was
where the paint-mixing analogy fell down. There was thus a need for some
explanation of how ancestral characteristics, hidden for one generation or more,
could reappear seemingly unchanged.

This need was met by the concept of **particulate inheritance.** There were
several particulate theories current in the late nineteenth century, but after
1900 only one of these, that proposed by the Austrian monk Gregor Mendel in
1865, proved useful in understanding the hereditary process in a wide variety
of organisms. All particulate theories of heredity maintain that parent organisms

pass on definite units or particles, each of which determine one characteristic of the offspring (flower color, hair color, bone size, etc.). The mendelian concept holds that the offspring inherits two particles for each characteristic, one particle from each parent. If the two particles are different in the form of the characteristic (i.e., one determining red flowers, the other white flowers) there can be two possible results. In the case of morning glories, both particles (which we would refer to today as **genes**) show their effects. The offspring is correspondingly midway between the two parents and is a true blend for that characteristic. A different result is seen in the case of eye color in human beings; here, blending does not occur. If a blue-eyed and brown-eyed set of parents have children, all will probably show brown eyes. However, sometimes brown-eyed parents have blue-eyed children, but blue-eyed parents never have brown-eyed children. According to the mendelian theory, the gene for brown eyes is said to be **dominant** over the gene for blue eyes. The latter, in turn, is said to be **recessive** to the gene for brown eyes. Thus a person with blue eyes is known to have only the genes for blue eyes. A person with brown eyes, however, may have two genes for brown, or one for brown and one for blue. Since the gene for brown eyes masks the effects of the gene for blue eyes, the latter is not expressed in the appearance of the organism. Yet, as in the case of the second-generation morning glories, the blue-eye gene can reappear in subsequent generations, apparently unaltered by its association with the dominant brown gene.

The mendelian concept of heredity is especially capable of explaining these inheritance patterns, where the concept of blending inheritance is wholly inadequate. According to the mendelian scheme, each gene for a certain characteristic is unchanged by its association with other genes. Thus, in the case of morning glories, the gene for "red flower" is unaltered by existing next to a gene for "white flower" in the first-generation organisms. When these organisms are crossed, some offspring of the second generation get two "red" genes (thus showing only red flowers), some get a "red" and a "white" gene (thus showing the pink condition) and still others receive two "white" genes (thus showing the white flower). Thus, although the first generation offspring showed neither red nor white flowers, the genes, when properly matched up in the second generation, produced the original parental characteristics again.

Mendel's particulate concept not only offers a more complete explanation of heredity than the idea of blending inheritance, but it also affords the means of predicting with considerable accuracy the results of crosses between a variety of parental types. For these reasons, Mendel's theory gained wide acceptance among biologists in the twentieth century. It is basically this theory, expanded through new work in molecular biology, which forms the substance of the modern-day view of heredity.

Acceptance of a particulate theory of inheritance raises many questions. What is the actual nature of the genes? How are these units passed on to the next generation? How does a gene determine, physiologically, a given characteristic? How do variations in the organism arise? These and many other questions have provided the stimulus for much genetic research in the present century. In the past decade, remarkable progress has been made in understanding the transmission and action of genetic material.

2-4 THE MUTATION THEORY

One of the important features of the modern theory of inheritance is that it explains how organisms produce nearly perfect copies of themselves. However, variations sometimes do arise in offspring. That is, there appear to be mistakes in the process by which genes reproduce themselves, in passing from one generation to the next. Such mistakes, if they cannot be shown to arise in accordance with mendelian rules, are said to be **mutations.** Mutations are due to changes in the hereditary code comprising the gene itself. Such changes generally show up as differences in the physical or chemical structure of an organism. A condition known as vestigial wing in the fruit fly is the result of a mutation of the gene for normal wing size. Flies bearing the mutant gene have very short, useless wings (see Fig. 2–3). Most mutations produce only very slight changes in the organism —so slight that only an expert can detect their presence. Occasionally, however, as in the vestigial-wing example, mutations may be quite noticeable.

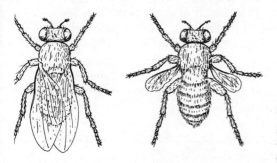

FIG. 2–3. The fruit fly *Drosophiia melanogaster.* The drawing to the left shows a normal or wild individual. To the right can be seen a mutant form, vestigial wing. Note the short, shriveled wings of the mutant.

Once mutations occur, they are generally passed on to subsequent generations. That is, a mutant gene "breeds true" until such time as it mutates further. Biologists do not know exactly what causes mutations to occur under normal conditions of life. However, there are various experimental means of speeding up the rate of mutation. X-radiation or chemical compounds such as mustard gas will do this very effectively. Such experiments indicate that mutations occur at random. Thus, although it is possible to increase the rate at which mutations occur, it is not possible to cause a specific gene to mutate by applying any of the above experimental methods.

2-5 THE CONTINUITY OF THE GERM PLASM

In 1889 the German biologist August Weismann formulated the theory of the "continuity of the germ plasm." In this theory, Weismann postulated that from the very first divisions of the fertilized egg, certain cells are "set aside" to become the reproductive or gamete-forming tissue of the developing organism. These he referred to as germ cells, or the **germ plasm.** The other cells develop into body cells, or **somatoplasm** (muscle cells, nerve cells, skin cells, etc.). The germ plasm was thus viewed as distinct from the somatoplasm from the very beginning of embryonic development. The germ plasm of the offspring descends directly from the germ plasm of the parent.

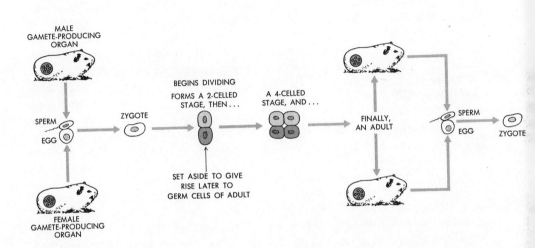

FIG. 2–4. Weismann's concept of the continuity of the germ plasm. Note that from the first division of the zygote, certain cells are set aside to become the future germ plasm which will later form sperm or eggs.

Weismann's theory provides a way of looking at the inheritance of various characteristics. It shows an unbroken continuity existing between the germ plasm of the parent and the germ plasm of the offspring. The somatic cells can be viewed as the agent involved in maintaining and transmitting the germ plasm. Weismann's concept extended the continuity which Virchow saw (continuity from one *cell* generation to the next) to include generations of entire organisms (see Fig. 2–4). It must be pointed out that Weismann's generalization is limited to the animal world only. No such separation of germ plasm and somatoplasm occurs in the embryonic development of plants.

Weismann's work indicated that in animals, at least, *acquired characteristics are not inherited.* Only changes which affect the genetic material in the germ plasm of an organism will be transmitted to the offspring. Thus, for example, any kind of mutation that may occur to genes in skin tissue will not affect the offspring of that individual. Mutations in the genetic code of gamete-forming tissue will, however, be passed on to the next generation.

2-6 EVOLUTION AND NATURAL SELECTION

The concept of evolution and of the means by which it occurs is probably the most important generalization in modern biology. Our present theory of the evolution of plants and animals consists of two conceptually and historically different ideas. One is the idea of evolution itself—the view that the many species of plants and animals existing today have descended, through modification, from earlier, different species. The other is the more theoretical consideration of natural selection—the mechanism by which this development is brought about.

The theory of evolution is an attempt to explain the diversity of animal and plant species. Modern evolutionary theory holds that groups of organisms change in the course of geological time. Species living today have descended, in modified form, from species which existed millions of years ago. The modifications which occur from one generation to the next are in general very small. Once these modifications have occurred a first time, however, they are passed on to succeeding generations unless canceled by other variations. After thousands or millions of years, accumulation of these slight modifications results in an animal or plant significantly different from its remote ancestors.

The theory of natural selection describes a *mechanism* by which evolution may occur. The concept of natural selection is closely linked with the name of Charles Darwin (1809–1882). It was primarily through Darwin's work that the idea of natural selection was firmly established as an important biological generalization.

The idea of natural selection as a mechanism for evolution is based upon several ideas:

1) Hereditary differences occur among members of a given species. Some of the variations will increase the individual organism's chances of survival, while others will decrease the chances.

2) More organisms are generally born than can survive in view of limited food supply, available habitats, etc.

3) Limited availability of the necessities of life produces *competition* among organisms. Competition is most severe between individuals of the same kind, because these have identical, or nearly identical, requirements.

4) Organisms which are successful in competition will have a better chance to reproduce.

5) The basic unit of evolution is the breeding population. This is a group of organisms of the same kind (i.e., the same species) who are capable of inter-

breeding. For example, a group of snowshoe hares inhabiting the same territory form a breeding population.

These are the ingredients of the natural-selection process. How do they produce evolution? Note first that those individuals whose hereditary variations enable them to get more food or better living conditions will survive longer and thus reproduce more. They thus pass on their genes to a larger percentage of the next generation than the less successful organisms. Over a period of many generations, then, the genetic character of the breeding population will change, favoring the more successful variations. Individuals in the population will gradually become better and better adapted to their environment.

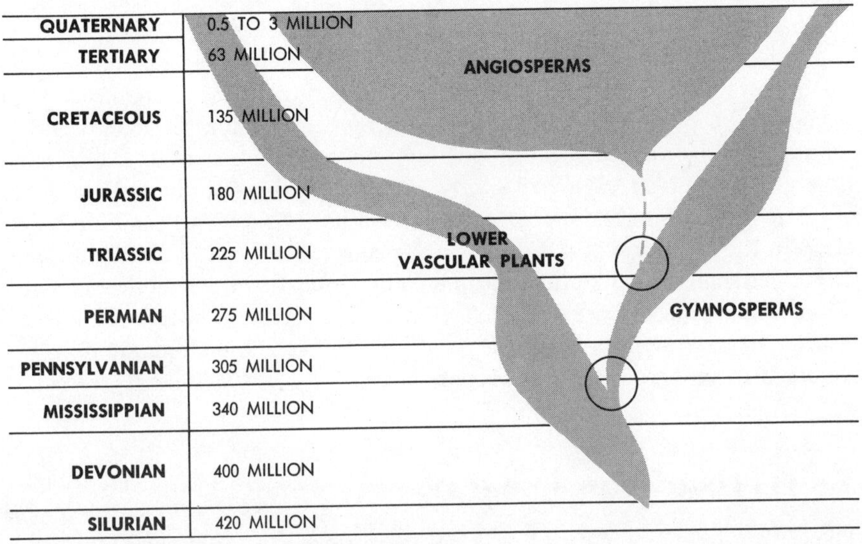

QUATERNARY	0.5 TO 3 MILLION		
TERTIARY	63 MILLION	ANGIOSPERMS	
CRETACEOUS	135 MILLION		
JURASSIC	180 MILLION		
TRIASSIC	225 MILLION	LOWER VASCULAR PLANTS	
PERMIAN	275 MILLION		GYMNOSPERMS
PENNSYLVANIAN	305 MILLION		
MISSISSIPPIAN	340 MILLION		
DEVONIAN	400 MILLION		
SILURIAN	420 MILLION		

FIG. 2–5. Phylogenetic chart of land plants. Various geological periods are shown on the scale to the left, with the oldest at the bottom. The figures beside each period indicate approximately how many years ago the period began. The two points in time where a common ancestor gave rise to two branches are circled. The width of the shaded area indicates the relative number of genera in each major branch at each point in time. (Adapted from *Scientific American*, February 1963, p. 81.)

The key to Darwin's concept of evolution is the idea of **differential fertility.** This simply means that inherited variations are judged as *successful* or *unsuccessful* according to whether they increase or decrease the reproductive capacity of the organism. Those organisms are successful which are able to produce a large number of offspring. Thus longevity is a criterion of success only insofar as it allows an organism more opportunity to reproduce. In darwinian terms, many species of insects are very successful even though as adults they may live for only a few days, for during this time they are able to produce a large number of offspring.

The concept of natural selection is an extremely important one. It provides a single generalization by which the evolution of all known living forms can be explained. It is integrally connected to the ideas of heredity, ecology, taxonomy, cytology (the study of cells), paleontology (the study of fossils), embryology, and even biochemistry. It thus brings together a number of areas of biological thought.

Evolution by natural selection has resulted in the proliferation of many kinds of organisms on earth. There are many more species alive today than there were several million years ago. For example, the many varieties of dog living today probably evolved from a common ancestral type. Thus not only does evolution include the gradual modification of a given kind of organism into a new kind, but also the divergence of a single kind of organism into many. Darwin especially emphasized the divergence pattern of evolution.

A family tree, or **phylogenetic chart,** may be used to make clear the historical relationships between groups evolving from a common ancestor. Figure 2–5 shows the broad outline of a family tree for the evolution of land plants. The points representing common ancestors are circled. At these points divergence occurred, giving rise to two separate groups of plants in each case.

The term "missing link" has often been used to refer to some undiscovered stage in the evolution of a group of organisms. The term "link" is unfortunate. It implies something like the link in a chain—one direct connection in a straight sequence of objects or events. A common ancestor is not at all like a link. Rather, it represents a starting point from which two different groups have diverged to evolve in their own separate ways.

2-7 ECOLOGICAL INTERRELATIONS

Darwin's concept of natural selection emphasized the important interrelation between an organism and its environment. Survival depends on how well an organism copes with the various aspects of this environment. Detailed studies have shown that all living things in a given area are closely interrelated, both with each other and with the nonliving aspects of the environment. The environment of an organism is the sum total of its biological and physical surroundings. The biological environment of any organism includes the other plants and animals in the region; the physical environment includes such features as temperature, amount of light, moisture, minerals, and so on. The aspect of biology which studies the relations between an organism and its environment is known as **ecology.**

Organisms exist together in **communities.** The groups of organisms which inhabit the forest, the field, the pond, or the ocean bottom are all examples of biological communities. All communities in a given region (for example, in a valley between two mountain ranges) are interdependent. The phrase "web of life" has often been used to express this interdependence, because disturbance to any one aspect of the community interrelations will have profound effects on all other aspects, just as breaking a single strand in a delicately woven spider's web can totally alter the entire pattern of the web. Man has learned, often the hard way, that to disturb the balance of nature in any region can produce far-reaching effects.

Around the turn of the century, air pollution from the smoke of a copper smelter destroyed much of the vegetation in Copper Basin, Tennessee. This resulted in a change in annual rainfall* in the region, which in turn restricted the type of plants which could regain foothold. Even present attempts to reforest the area have been greatly hampered by continual erosion. This example illustrates in a dramatic way how closely the biological and physical environments are interrelated.

The generalization of ecological interrelations is of great practical as well as theoretical importance. It provides the basis for all efforts to conserve wildlife, preserve forests, and to make better use of land for agriculture. All sound conservation programs are based upon an intimate knowledge of the principles of ecology. Since conservation is one of the most crucial social and technological issues which mankind faces today, ecology is one of the most important areas of modern biological research.

2–8 THE MOLECULAR FOUNDATION OF BIOLOGY

One general tenet underlies much current biological research. This is the idea that ultimately, living processes can be explained in terms of the interaction of molecules, atoms, and subatomic particles such as electrons. This does not mean, of course, that research in areas of ecology or evolutionary theory must be concerned with molecular explanations in order to be important. We must always keep a concern for the whole organism. Yet current study in molecular biology indicates that areas such as genetics, evolution, general physiology, and even classification of organisms, can all be carried to the molecular level.

The molecular approach to biology is based on our present-day understanding of physics and chemistry, and their applicability to biological problems. As an area of current research, molecular biology is extremely important for a variety of reasons. One is that by using the tools of physics and chemistry, it is possible to treat many biological problems with precision and with rigor. For instance, through modern studies on the physics and chemistry of energy processes in living systems, it is possible to predict with great accuracy the amount of heat which an organism will release in performing a specific task. It is possible too, to calculate exactly how much energy is released in the breakdown of a given

* Rainfall in a region is affected to some extent by the amount of moisture released into the atmosphere locally by plants. The loss of plant life over an extensive area like Copper Basin could thus significantly affect the average rainfall in this region.

number of food molecules, or how much energy is used in building a certain complex molecule like a protein. Studies in molecular biology have also yielded a very precise indication of how hereditary information is coded so that the offspring duplicates the characteristics of the parental type. By using the methods of chemistry and physics, such principles have been rigorously established. Much less has been left to speculation or to guesswork than would otherwise be the case.

Another important reason for the central role of molecular biology in much current research is that such studies have revealed clearly how similar the chemical processes are in all organisms. The processes involved in energy transformation or in the passing on of coded heredity information seem to be quite similar in organisms as different as a maple tree, a bacterium, and man. In other words, molecular biology has brought to light the great biochemical unity which seems to underlie the living world. The fact that in quite different organisms the same molecules can be found performing the same chemical jobs indicates how much seemingly different organisms have in common. It gives strong support to the idea that all organisms have descended ultimately from a common ancestor.

2-9 MECHANISM AND VITALISM

There may be some doubt as to whether the ideas of "mechanism" and "vitalism," as ways of looking at living organisms, are generalizations in the same sense as those of natural selection or the gene concept. Some people might argue that they are only philosophies. Clearly however, they represent opposing views of what is an acceptable biological explanation of the nature of life. Although debated more hotly in the past than at the present, their basic assumptions strike more closely to the problem of "What is life?" than perhaps any other biological issue.

Mechanism is that view of living organisms which holds that life is completely explicable in physical and chemical terms. A mechanist believes that organisms are extremely complex organizations of various atoms and molecules. He also believes that the interrelation between these atoms and molecules can be understood by applying the tools of the chemist or physicist. He does not believe that there is anything to a living cell that goes beyond the physical or chemical concepts which we accept today. He is willing to accept the fact that modern techniques are not perfect, and that newer methods will have to be developed before we can completely understand how a living cell functions. The important point is that to him, the possibility of gaining such an understanding is a real one.

Vitalism, on the other hand, maintains that life is an expression of something more than the mere interaction of a group of molecules. To the vitalist, the living organism is more than the sum of its parts. There resides in the living matter some vital principle which cannot be explained in physical or chemical terms. The vitalist does not deny that chemical analyses of living organisms are important. But he feels that "life" involves something more than the principles of physics and chemistry.

It is essentially over this question of whether or not a "something more" actually exists that the mechanists and the vitalists have argued for so long. Much debate by a number of scientists in the later nineteenth and early twentieth centuries did very little to settle the issue. In any case, the point seems largely academic. In their daily research, most biologists today are mechanists. They look for biological explanations which agree with those of physics and chemistry. Yet, such a biologist could also believe (and some do) that this approach will never provide a complete understanding of the living organism. They would maintain that there is something beyond this which will always defy man's laboratory analysis.

2-10 CONCLUSION

The generalizations discussed in this chapter have two major functions for the biologist. First, they provide a means by which he can relate many individual pieces of information. Second, they allow him to make predictions. As will become clear in the next several chapters, the predictive value of scientific generalizations is of fundamental importance.

Of the generalizations which we have discussed, some offer a higher level of generality than others; that is, they cover a larger number of cases than those with a low level of generality. Darwin's idea of natural selection is a high-level generalization because it applies to all known organisms. Plants, animals, bacteria, even viruses, evolve over time in accordance with Darwin's idea. Weismann's concept of continuity of the germ plasm, however, is a lower-level generalization because it applies only to animals. Thus not all of the generalizations discussed in this chapter have the same weight, or the same value, to the biologist. More will be said about the nature and validity of scientific generalizations in the following chapter.

EXERCISES

1. What is the cell concept as we understand it today? What did the following men contribute to the development of that concept: (a) Matthais Schleiden? (c) Theodor Schwann? (c) Rudolf Virchow?
2. Describe the difference between the concepts of blending inheritance and particulate inheritance.
3. What is a mutation?
4. Briefly outline Darwin's conception of natural selection. How is the idea of mutations related to that of natural selection?

5. What is the distinction between germ plasm and somatoplasm?

6. What is meant by the phrase "continuity of the germ plasm"?

7. In darwinian terms, what is the criterion of "success" in any species?

8. Read the viewpoints of a mechanist and a vitalist as given in the suggested readings. Which arguments seem most valid? What loopholes, if any, do you find in each type of argument?

9. What is meant by saying that all biological processes have a molecular basis? What support is there for this assertion?

SUGGESTED READINGS

The mechanist-vitalist controversy represents some attempts to arrive at a philosophy of biology. Two books, each expressing quite different views on the subject, will provide the basis for stimulating thought about the nature and limitations of biological investigation.

DRIESCH, HANS, *The Science and Philosophy of the Organism* (London: Adam and Charles Black, 1908). Driesch was the most prominent and vocal of the late nineteenth- and early twentieth-century vitalists. The experiments which led him to his vitalism are described in this book.

LOEB, JACQUES, *The Mechanistic Conception of Life* (Cambridge: Harvard University Press, 1964). This is a reprint of a book which originally appeared in 1912. Loeb represents the extreme mechanistic outlook. The book is not too technical to provide a thorough introduction to the mechanist's arguments.

CHAPTER 3

THE NATURE AND LOGIC OF SCIENCE

3-1 INTRODUCTION

First and foremost, biology is a science, the biologist a scientist. Science can be distinguished from other fields of intellectual endeavor by two main features. Obviously, it differs in its content—the type of organized knowledge with which it is concerned. A more important difference, however, lies in the *procedure* of science—its strictly empirical approach to problems. Science deals only with rational beliefs which can be verified or disproved by observation or experiment. In the words of Roger Bacon (1210–1292), "experimental science has one great prerogative . . . that it investigates its conclusions by experience."

Science is frequently said to consist of collecting and organizing facts. This, however, is only one aspect of science. Far more important is what the scientist does with the facts he has at hand. The way in which the scientist draws conclusions, makes generalizations, and tests predictions forms the method of science. In a general way, then, it is useful to make a distinction between content and procedure in science. Scientific content is the subject matter of science—the generalizations which the scientific community may recognize as valid. The concepts embodied in Mendel's laws, the concepts of natural selection and mutation—these represent scientific content. The methods by which such concepts were obtained—by experiments, observations, or reasoning from known examples—represent scientific procedure.

17

3-2 THE "SCIENTIFIC METHOD"

The popular, man-on-the-street concept of the scientist and his methods is a poor one. According to this concept, the scientist is a man with secret means of obtaining knowledge to benefit mankind. The fact that explanations put forth by research scientists may be wrong as often as they are right, and that not all of their discoveries directly benefit man (indeed, many seem to be completely useless), is not widely known. This is possibly due to the fact that wrong guesses are not given much publicity, and right guesses which do not directly benefit man attract less public attention.

Even among scientists, however, there is wide disagreement as to what is meant by the "scientific method." Some science textbooks list a series of six or seven steps involved in the scientific method. Such a formal and highly structured description is quite unrealistic. No research scientist follows any such formalized ritual in performing his experiments.

Some writers, however, have gone to the other extreme in their description of the scientific method. One states that "science is simply doing one's damnedest with one's mind with no holds barred." This view has one positive attribute. It correctly indicates that the means used by scientists in solving problems are not necessarily unique to science. As a definition of the scientific method, however, the statement is not very successful. Followed to its logical conclusion, it indicates that philosophers, mechanics, mathematicians, plumbers, or any other persons who work diligently to solve problems are also scientists. Most certainly this is not the case.

Science proceeds by postulating and testing hypotheses. **Hypotheses** are simply tentative explanations put forth to account for observed phenomena.

Let us take a specific example:

The silver salmon, *Oncorhyncus kisutch*, hatches in the freshwater streams of our Pacific Northwest. The young fish swim downstream to the Pacific Ocean, where they may spend five years attaining full size and sexual maturity. Then, in response to some undetermined stimulus, they return to fresh water to spawn (lay their eggs).

By tagging the fish, a remarkable fact is discovered. Nearly always, *the fish return to the precise stream where they were born.*

Here is an observed phenomenon which arouses the curiosity. *How* are the fish able to locate the precise stream in which they were born? This is no easy task. Some of the fish must swim up high waterfalls and go as far inland as the state of Idaho in order to return to their place of birth.

An hypothesis is needed to explain the phenomenon. In a sense, an hypothesis is simply an "educated guess." In this case, the hypothesis will probably be based on pertinent observations of the salmon and their habits. Perhaps the fish find their way to their homes by recognizing certain objects they saw when they passed downstream as young fish on their way to the sea. Or perhaps they recognize the "taste" or "odor" of their home streams. Several other hypotheses are possible, of course, but let us settle on these two. What next?

Clearly, to stop here, after merely formulating hypotheses, is not very satisfying. It is natural to want to find out which hypothesis (if either) is correct. The scientist, in his attempts to find the answer, proceeds by designing and performing experiments. *The primary purpose of scientific experiments is to test hypotheses.* Thus any hypothesis selected by a scientist to explain a natural phenomenon must meet a very important requirement: *It must be testable.*

Both hypotheses which we selected to explain the homing behavior of salmon meet this requirement; they *can* be tested by experimentation. But *how* do experiments test hypotheses? The answer is quite simple. *Experiments test hypotheses by testing the correctness of the predictions that can be derived from them.*

FIG. 3–1. This "truth table" shows the relation between an hypothesis and its predictions. Note that a true prediction may be derived from a false hypothesis as well as from a true one. Thus true predictions do not constitute proof of the truth of an hypothesis.

A TRUTH TABLE

HYPOTHESIS	CONCLUSION OR PREDICTION
TRUE	TRUE
FALSE	TRUE or FALSE

Consider, for example, the first hypothesis, which explained the salmon's ability to find their home stream solely on the basis of visual recognition. If this hypothesis is correct, then salmon with shields placed over their eyes should be unable to find their way home. This reasoning can be expressed more formally as follows:

Hypothesis: If ... *Oncorhyncus kisutch* salmon use visual stimuli alone to find their way to their home streams to spawn, ...

Prediction: then ... blindfolded salmon of this species should not be able to find their way home.

Suppose that the fish find their way home when blindfolded just as well as they did before. If we assume that no other factors (or variables) have been overlooked which might influence the results, can it be stated that the experimental results *disprove* our hypothesis? Yes. Suppose, on the other hand, that the blindfolded fish did *not* find their way to their home streams. Would these results *prove* the visual-stimulus hypothesis? No. The experimental results can only be said to *support* the hypothesis.

This raises an interesting question. Why should it be possible to *disprove* an hypothesis by one set of experimental results, yet not be able to *prove* the same hypothesis by obtaining the predicted set of results? The answer lies in the nature of the relation between hypotheses and the predictions which can be derived from them. This relation, which is shown in the "truth table" of Fig. 3–1, forms the basic framework for the operations of **deductive logic.**

Deductive logic (often called *if ..., then* reasoning) is the heart and soul of mathematics. It becomes most evident in plane geometry, e.g., "*If* two points of a line lie in a plane, *then* the line lies in the same plane." However, deduction plays no less of a role in other fields of mathematics as well, e.g., "*If $a < b$ and $x \le y$, then $a + x < b + y$*" (the addition law), or "*If $x < y$ and $a > 0$, then $ax < ay$*" (the multiplication law).

In science, and therefore in biology, deduction is just as vital as it is in mathematics. (Recall our breakdown of the salmon experiment into an *if . . .*, *then* framework.) Yet, there are important differences between the way deduction is used in mathematics and the way it is used in experimental science. Mathematicians generally deal with symbols. They are not so concerned with physical entities such as migrating salmon. Furthermore, the mathematician can manipulate his symbols at will. He can create situations in his proofs in which he is certain that only one hypothesis is being tested, only one question being asked. Not so the biologist. The salmon he is studying cannot be so easily manipulated. Therefore, the biologist can never be absolutely certain that his experiment has eliminated all the variables which might influence his results. Shielding the eyes of the salmon, for example, may cause the animals to use

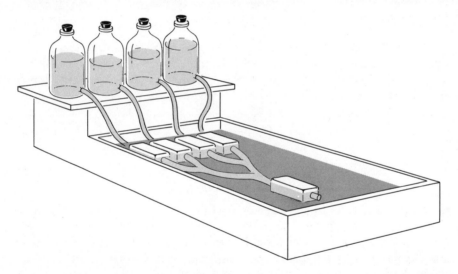

FIG. 3–2. Whenever possible, the biologist attempts to construct an experimental apparatus which allows him to test his hypotheses in the laboratory, rather than in the field. In this way, variables which might influence his results can be more easily controlled. This apparatus tests the ability of young eels to detect minute quantities of dissolved substances in the water. The object of the experiment is to discover whether these substances play a role in the eels' instinctive ability to find their way home from the Sargasso Sea (a region in the Atlantic Ocean) where they were born to the freshwater streams where they spend most of their lives. In this experiment, it was shown that the eels had no preference for tap water over seawater, but definitely preferred natural inland water to seawater. They were able to detect certain organic substances in the water even when these were diluted to 3×10^{-20} parts per million. This means that the eels must be reacting to the presence of only two or three molecules of such a substance in their scent-detecting olfactory sacs.

another sensory system in order to find their way home. Perhaps, normally, they *do* use their eyes to find their way home. Such a possibility may seem far-fetched, and highly unlikely. In the case of the salmon, it probably is. But the fact that such things are even remotely possible must constantly be in back of the biologist's mind.

A major problem in biological research, then, becomes one of experimental design (to which considerable attention will be devoted in Chapter 4). The biologist recognizes the impossibility of eliminating all the variables which might affect his experimental results, but he tries to design his experiments to decrease the *likelihood* that these variables will occur (see Fig. 3–2).

Examine again the truth table in Fig. 3–1. Note that the word "conclusion" as used by a mathematician is interchangeable with the word "prediction" as used by the biologist, for the predictions which can be made from an hypothesis are simply the conclusions that one must draw from accepting it. In the case of the salmon, it must be concluded (or predicted) that the blindfolded salmon could not find their way home if the visual-stimulus hypothesis is accepted as being correct. If blindfolded salmon *do* find their way home, our conclusion has proved false. From the truth table, we see that this automatically means that our hypothesis is false, for note that *a true hypothesis can never give rise to a false conclusion.* In other words, predictions derived from a true hypothesis should never lead to contradictions.

The truth table also shows that we can never *prove* that an hypothesis is true. For, while a true hypothesis always gives rise to true predictions, *so also may a false hypothesis.* The importance of this last fact cannot be over-emphasized, for it shows that science can only deal with its "truths" in terms of probabilities, and never in terms of certainties.

In the past, many false hypotheses have been held by scientists and laymen alike, simply because accurate predictions could be made from these hypotheses despite their falsity. Acceptance of the belief that the sun orbits the earth leads one to predict that the sun will rise on one horizon, cross over the sky, and set on the other horizon . . . and so it does.* The fact that this prediction turns out to be correct does not, of course, mean that the sun *does* orbit the earth. In order to demonstrate that this hypothesis is false, other tests must be devised which show it to yield false predictions.†

---

* It is likely that the observation of this aspect of the sun's behavior contributed to the idea that the sun orbits the earth. This illustrates the fact that hypotheses frequently arise from observations of the very phenomena which the hypotheses would predict. It is often difficult to establish which came first.

† One such test is to predict the future relative positions of the sun, earth, and other planets, given that the sun does orbit the earth. Such predictions are invariably shown to be false, thus disproving the hypothesis. On the other hand, accepting the hypothesis that the earth, along with the other planets, orbits the sun leads to very accurate predictions regarding the relative positions of the sun, earth, and other planets at any point in time. Every such accurate prediction supports the earth-orbiting-the-sun hypothesis.

Although the truth table shows that a true hypothesis never gives rise to a false conclusion (prediction), only in mathematics does the obtaining of just one false conclusion spell certain death for the hypothesis. Biologists rarely deal with cases in which *every* prediction made by an hypothesis turns out to be correct. The question then becomes one of *how many* or *what proportion* of a given number of predictions must be verified in order to make the hypothesis a useful one. For this reason, experimental data are often subjected to a **statistical analysis,** in which mathematics is employed to determine whether deviations from the pattern that is predicted by the hypothesis are significant. This topic will be dealt with more thoroughly in Chapter 5.

Let us return once more to the problem of explaining salmon homing behavior and test the second hypothesis, which proposes that the fish find their way back to their home stream by their sense of smell. This hypothesis is supported by a chemical analysis of the water in several different streams. Such an analysis shows the water of each stream to be distinctively different from that of any other, due primarily to the different kinds and quantities of dissolved minerals which each contains.

We can proceed, then, to test experimentally this second hypothesis by testing the validity of a prediction which can be made from it:

Hypothesis: *If . . .*   *Oncorhyncus kisutch* salmon find their way to their home stream by following its distinctive odor upstream, . . .

Prediction: *then . . .* blocking the olfactory sacs (with which the fish detect odors) should prevent the salmon from finding their home stream.

This experiment was performed by Dr. A. D. Hasler and his associates, of the University of Wisconsin. The results strongly supported the odor hypothesis; a large majority of the fish were unable to find their way to their home stream to spawn. Nevertheless, some fish *did* find their way. Each one that did represents a false prediction which, according to the truth table, proves the hypothesis to be false. However, the laws of probability must be taken into consideration here. Statistical analysis shows that a certain number of the fish would be expected to end up in their home streams purely by chance. Since the number of experimental fish which found their way home was not significantly greater than the number which would be predicted to get there by chance, the odor hypothesis can still be considered a valid one.

Let us take a second example of experimentation in biology to illustrate another point concerning the interpretation of experimental results:

It has been shown many times that exposure of certain strains of mice to X-ray beams of 600 roentgens or more (a roentgen is a unit of measure of the amount of energy delivered in X-ray beams) causes death within two weeks or less. The death seems to be due to secondary rather than primary effects of the radiation. But it is uncertain just what is the primary cause of death at any one time, especially in the period of one to five days after exposure. It was thought that death might possibly be due to bacterial infection resulting from a migration of bacteria through the intestinal epithelium (lining), which histological (tissue) examination showed had been severely damaged by the X-rays. In order to test this hypothesis, antibiotics of various types were administered to the irradiated mice in several different ways to see whether this had any effect on the time of death. However, no such effect was shown, as the mice still died in the same length of time as the control animals, which had been irradiated under the same conditions but given no antibiotics. It was tentatively concluded, therefore, that death in the period tested (from one to five days after exposure) was not due to bacterial infection.*

Note the deductive *if . . .* , *then* reasoning here. The experimental logic can be simply stated as follows:

Hypothesis: *If . . .*   the deaths of irradiated mice within one to five days after exposure are due to bacterial infection, . . .

Prediction: *then . . .* administration of antibiotics should lower the death rate of mice which receive them.

The experimental results showed the prediction to be a false one. The mice still died in the same length of time after exposure to the X-rays. Thus we know, barring experimental error, that the hypothesis explaining the deaths as being due to bacterial infection was also false, and therefore must be either discarded or modified.

Suppose that the administration of antibiotics *had* caused a lengthening of life. Would this have shown that our hypothesis must be the correct explanation? Absolutely not, although this result would have lent strong support to the *probability* of its being correct.

Could it be stated that death from radiation *in animals* is not due to bacterial infection? No, for the word "animals" includes many more forms of life than just mice. Could it be stated that death from radiation *in mice* is not due to bacterial infection? No, for not all strains of mice were tested. When the research paper is written for publication in a research journal, the biologist will carefully word his interpretations, limiting them to the precise strains of mice tested and to the time period of death (one to five days after exposure) with which he worked.

Despite the limitational care with which experimental results are generally interpreted, biologists often *do* extend their experimental results from one organism to another. Modern medical drugs, for example, are usually tested first on laboratory animals; if successful, their use may be extended to humans. But there is always an element of uncertainty involved. All organisms do not necessarily react the same way to the same drugs . . . "One man's meat is another man's poison."

---

* J. J. W. Baker, Roscoe B. Jackson Memorial Laboratory, 1960, unpublished notes.

3-3 INDUCTIVE LOGIC

All the generalizations discussed in Chapter 2 are based on observations and/or experiments extending, in many cases, over a considerable number of years. The cell concept, Mendel's ideas of heredity, and Darwin's theory of natural selection are generalizations drawn from observations made on many different organisms. They are **inductive generalizations** attained through a process of **inductive logic**.

Inductive logic involves coming to a probable conclusion on the basis of many particular instances. Suppose, for example, that a person tastes a green apple and finds that it is sour. He tastes a second green apple; it also is sour. A third and fourth green apple yield the same results. From these separate, individual observations, a general conclusion might be drawn: "All green apples are sour." Inductive logic, then, involves proceeding from the specific to the general. In this case, it involves going from specific observations on four green apples, to a general conclusion about *all* green apples. Inductive logic is therefore an opposite of deductive logic, for the latter proceeds from the general to the specific.

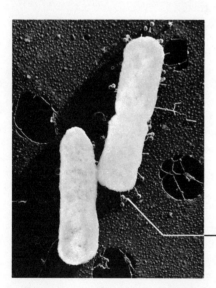

FIG. 3-3. The bacterial cell at the left shows several bacteriophages attached to it. Some bacteria, however, are immune to them. Luria and Delbrück performed an experiment to determine whether the mutation causing immunity was a random one or was due to exposure to the phages (see text for discussion). (From *Scientific American*, June 1961. Courtesy Thomas F. Anderson, Francois Jacob, Elie Wollman.)

BACTERIOPHAGE

Although inductive and deductive logic are two distinct types of thought processes, scientists do not tend to think exclusively in one way or the other at different times. In the solution of scientific problems there is constant interplay between inductive and deductive thinking. Consider again the generalization, "All green apples are sour." After tasting the first apple, a person might conclude that only this one apple was sour. After tasting the second, however, he might immediately conclude that *every* green apple is sour. It might be said that this is "jumping to conclusions." In more precise terms, it is simply making an inductive generalization on the basis of only two items of information. From this generalization, a prediction can be made:

If all green apples are sour, . . .*

then, the next green apple tasted should be sour.

A quick test confirms the prediction. This, in turn, lends support to the original inductive generalization. Thus we see a constant interplay between deductive and inductive reasoning in problem solving.

From the above example it is evident that the more observations are available, the more reliable are the inductive generalizations which can be drawn from them. An inductive generalization based on two items of information is less likely to be reliable than one based on ten or a hundred. However, inductive generalizations never attain absolute certainty. They only attain high degrees of probability. The degree of certainty attained depends on both the amount and the kind of information used in drawing the generalization. There are some apple varieties, for example, which are both sweet and green. Should the apple-taster encounter such apples, it would be necessary for him to modify the initial generalization accordingly, i.e.: "All green apples are sour, except for variety X."

The following experiment illustrates the interplay of inductive and deductive logic which often occurs in science. The experiment was designed to answer the question of how mutations arise in living organisms. For many years, some biologists maintained that mutations (see Section 2–4) were triggered within the organism by specific changes in its environment. Other biologists disagreed. They felt that mutations occurred entirely at random, and were quite independent of environmental influences. The conflict existing between these views may be expressed in the form of a question: Do mutations arise spontaneously on their own, or are they due to environmental influences? In 1943, using bacteria as their experimental organism, the geneticists Luria and Delbrück performed an experiment which answered this question.

Bacteria are parasitized by certain viruses called **bacteriophages.** Figure 3–3 shows an electron micrograph of several such bacteriophages (called "phages" for short). If a culture of bacteria is allowed to grow for several days and then exposed to phages, most of the cells are killed. A few may survive, however. These survivors are "resistant" varieties. Their resistance is passed on to their descendants; it is, therefore, the result of a genetic mutation.

On the basis of this information, two working hypotheses can be formulated:

1) The resistance to phages arises in the bacteria by spontaneous mutation. Such mutant bacteria will appear whether or not the bacterial culture is exposed to phages. In the absence of the selecting agent (the phages), resistant cells are simply not detected among the greater masses of nonmutant bacterial cells. When phages are introduced, however, only the mutant forms survive and reproduce.

---

* Note that despite the deductive format, the statement, "All green apples are sour," is not the same type of hypothesis as "Salmon find their home streams by sense of smell." The latter hypothesis attempts to *explain* a phenomenon, the former, simply to summarize. Both types of hypothesis occur in science.

2) The resistance is induced in some of the bacteria by their contact with the phages. The bacteria which respond to this change in environment (i.e., the introducing of the phages) by mutating, survive. The bacteria which do not respond are destroyed.

In other words, the second hypothesis holds that the presence of the phages is the causal agent for mutation and that mutants for phage resistance do not appear until the bacteria come in contact with the phages. By contrast, the first hypothesis holds that the mutants are present all along but are simply not detectable until the phages are introduced.

To test the first hypothesis, Luria and Delbrück set up a number of bacterial cultures of the same species. Each culture was grown from a small group of bacterial cells. All the bacterial cultures were simultaneously exposed to phages, and the number of resistant cells, or survivors, were counted.*

The experimenters reasoned in the following way.

Hypothesis I: *If* ...   mutations occur spontaneously, ...

Prediction: *then* ... the number of resistant cells in the various culture dishes should be quite different.

If, for example, the mutation occurs early, when the growing culture contains few cells, the mutant cell will multiply and leave a large number of offspring each bearing the mutation. By the time the phages are introduced into the culture, many resistant bacteria may be present. Conversely, if the mutation takes place just before the introduction of phages, only a few resistant cells will be present. The laws of chance predict that there would be considerable variation in the number of surviving cells per culture dish.

Hypothesis II: *If* ...   mutations occur in response to the presence of phages, ...

Prediction: *then* ... the number of resistant cells per culture dish should be quite uniform.

Since each culture dish contains roughly the same number of bacteria, and the amount of phages introduced in each case is the same, hypothesis II predicts that the number of mutants should be roughly the same from one culture to the next.

The results of this experiment (Fig. 3–4) show that the variation in number of surviving cells per culture dish is quite large. Some cultures have only two or three surviving colonies, while others have twelve, fifteen, or more. The experimental results bear out the prediction of hypothesis I. Thus hypothesis I can be said to be supported. The results contradict the prediction of hypothesis II. Thus, barring experimental error, hypothesis II can be said to be disproved.

---

* The number of survivor cells can be counted quite easily by allowing the culture to incubate for a few days. Each surviving cell will reproduce to form a colony which can be detected by simply examining the culture dish.

Mutations *do* occur at random, and are entirely independent of the environ-mental changes which may give them selective value. Note that the Luria and Delbrück experiment can be said to test both hypotheses. Since the pre-dictions made by hypotheses I and II are contradictory, the results can support only one of them. Such an experiment is called a **crucial** experiment. Another example of a crucial experiment will be discussed in Chapter 4.

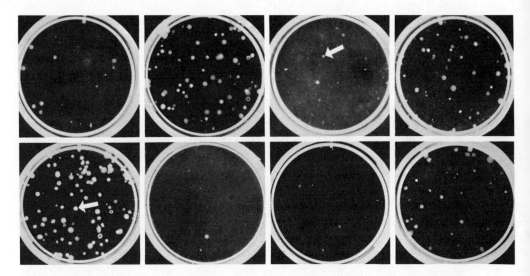

FIG. 3–4. The random variation in bacterial colony numbers obtained in the Luria and Delbrück experiments clearly supports the hypothesis proposing that mutations occur spontaneously and at random (see text for discussion).

In the case just discussed, the experimental results overthrew a rather widely held concept. Most of the time, however, the results are not so dramatic. Far more often, experimental results either support already well-established hypotheses, or cause minor modifications of them. For example, simple experi-ments on the respiratory physiology of a tree, a man, a robin, a frog, and a fish might disclose that the simple sugar glucose is oxidized as energy is released. At the same time, oxygen is used up, and carbon dioxide and water are released as waste products. From these facts, one might propose hypothesis I: "The energy needed by organisms is obtained from the oxidation of glucose by oxygen." In summary form, we may write the balanced equation as follows:

$$C_6H_{12}O_6 + 6O_2 \rightarrow 6CO_2 + 6H_2O.$$

If no further work is done, hypothesis I might attain the status of a full-fledged theory or law. But investigations that have been carried out on the cells of other organisms, such as yeast, disclose that these cells can tap the energy of glucose without oxygen. Another oxidizing agent (pyruvic acid) is used. Thus our original hypothesis I must be modified to read: "The energy needed by organisms is obtained from glucose by some oxidizing agent." The discovery that some organisms use substances other than glucose as the source of their energy leads to further modification of hypothesis I: "The energy needed by organisms is obtained from energy-rich substances by oxidation of these substances."

Thus each new discovery necessitates new modifications of previously acceptable hypotheses. The whole apple is not discarded; its bad parts are merely pared away. Because of many modifications and additions our current hypotheses regarding respiratory physiology are considerably more refined than those discussed here.

In a very real sense, then, scientific "truths" are only *approximate* truths. In the words of the chemist G. N. Lewis:

The scientist is a practical man and his are practical aims. He does not seek the **ultimate** *but the* **proximate**. *He does not speak of the last analysis but rather of the next approximation. His are not those beautiful structures so delicately designed that a single flaw may cause the collapse of the whole. The scientist builds slowly and with a gross but solid kind of masonry. If dissatisfied with any of his work, even if it be near the very foundations, he can replace that part without damage to the remainder. On the whole, he is satisfied with his work, for while science may never be wholly right it is certainly never wholly wrong; and it seems to be improving from decade to decade.*

It is true that a deductive-logic framework is present in every scientific experiment, but this does not mean that every research scientist is constantly examining his experiments to make sure this framework is present. Rather, the deductive-logic framework is there because the designer of the experiment is a logical person, accustomed to thinking this way when designing experiments. In the laboratory, it is a kind of "second nature" to him.

It should be noted that many biologists never carry out scientific experiments in the full sense of the word. These biologists concern themselves with the gathering of factual material. The detailed anatomical examination of a new plant or animal species is an example of such work. No hypothesis is being tested here; no predictions are being made. Yet, such work is often of great value. It provides the factual tools which others may use to design and carry out significant experiments on the organisms involved.

3-4 THE APPLICATION OF LOGIC: A CASE STUDY

It is now known that a fluid called **semen,** produced by the males of higher animals, contains **spermatozoa,** or sperm. Sperm are living cells, consisting of a headpiece and a tail. They carry the inheritance factors of the male, and are capable of independent movement. In sexual reproduction, the sperm swim toward the female egg cell and unite with it to achieve fertilization. It is fertilization which causes the egg to begin its development into a new animal.

In the eighteenth century, however, scientists were still uncertain as to just *how* the male semen caused fertilization of the egg. The importance of the spermatozoa was not recognized. Thus only two possibilities were considered:

1) The seminal fluid of the male must make actual contact with the egg before it would begin development; or

2) It was only necessary that a gas or vapor, arising from the semen by evaporation, make contact with the egg.

From their examination of the human female reproductive system, physicians saw that the semen would be deposited a considerable distance from the egg. Since the role played by the spermatozoa was not recognized, the fact that they might be able to swim toward the egg was not taken into account. It was therefore assumed that only a vapor diffusing from the semen could possibly reach the egg to fertilize it.

On the basis of these anatomical observations, the vapor hypothesis gained considerable support. In 1785, it was put to experimental testing by the Italian, Lazaro Spallanzani (1729–1799).

The following examination of excerpts from Spallanzani's report analyzes his experiments and conclusions and demonstrates their logical basis.

Is fertilization affected by the spermatic vapor?
It has been disputed for a long time and it is still being argued whether the visible and coarser parts of the semen serve in the fecundation of (i.e., here, in triggering the development of) man and animals, or whether a very subtle part, a vapor which emanates therefrom and which is called the *aura spermatica*, suffices for this function.

Here the problem is defined: Does the semen itself cause the egg to develop? Or, is it merely the vapor arising from the semen that does so?

It cannot be denied that doctors and physiologists defend this last view, and are persuaded in this more by an apparent necessity than by reason or experiments.

The lack of experimental evidence to support the vapor hypothesis is pointed out here by Spallanzani. In the full text of his report, he cites some of the anatomical observations noted in the introductory part of this section.

Despite these reasons, many other authors hold the contrary opinion, and believe that fertilization is accomplished by means of the material part of the semen.

The alternative hypothesis—that the semen must actually make contact with the egg—is stated.

These reasons advanced for and against do not appear to me to resolve the question; for it has not been demonstrated that the spermatic vapor itself arrives at the ovaries, just as it is not clear whether the material part of the semen that arrives at the ovaries, and not the vaporous part of the semen, is responsible for fertilization.

The statement, "... it has not been demonstrated that ... (etc.)" again shows Spallanzani's recognition of the lack of concrete evidence to support either hypothesis.

Therefore, in order to decide the question, it is important to employ a convenient means to separate the vapor from the body of the semen and to do this in such a way that the embryos are more or less enveloped by the vapor;

An experimental design is suggested. Some sort of experimental apparatus must be constructed to properly answer the questions to be posed by the experiments.

for *if* they are born, [*then*] this would be evidence that the seminal vapor has been able to fertilize them; or [*if*] on the other hand, they might not be born, *then* it will be equally sure that the spermatic vapor alone is insufficient and that the additional action of the material part of the sperm is necessary.

Note the two occurrences here of the "if ..., then" format as Spallanzani cites the deductive basis of his experimentation.

(*Note:* Spallanzani had shown earlier that the semen could be diluted several times, yet still remain capable of fertilization. In terms of what is known today regarding the role of the spermatozoa in fertilization, this is not surprising. However, Spallanzani interpreted these results as support for the vapor hypothesis, since he considered the vapor to be merely diluted semen. The following experiment, however, convinced him otherwise.)

In order to bathe tadpoles [eggs]* thoroughly with this spermatic vapor, I put into a watch glass a little less than 11 grains of seminal liquid from several toads. Into a similar glass, but a little smaller, I placed 26 tadpoles [eggs] which, because of the viscosity of the jelly, were tightly attached to the con-

Spallanzani here describes his experimental apparatus (see Fig. 3–5). Often an important part of an experiment is the design of such apparatus.

---

* Like many men of his day, Spallanzani believed that the animal egg contained a tiny miniature of the adult form, which needed only fertilization to grow to full size. Hence his reference to the unfertilized eggs as tadpoles. The belief in the existence of a preformed individual in the egg (the *preformation theory*) will be discussed in Chapter 14.

cave part of the glass. I placed the second glass on the first, and they remained united thus during five hours in my room where the temperature was 18°. The drop of seminal liquid was placed precisely under the eggs, which must have been completely bathed by the spermatic vapor that arose; the more so since the distance between the eggs and the liquid was not more than 1 ligne [2.25 mm]. I examined these eggs after five hours and I found them covered by a humid mist, which wet the finger with which one touched them; this was however only a portion of the semen, which had evaporated and diminished by a grain and a half. The eggs had therefore been bathed by a grain and a half of spermatic vapor; for it could not have escaped outside of the watch crystals since they fitted together very closely.

But in spite of this, the eggs, subsequently placed in water, perished.

Although the experiment overthrows the spermatic vapor theory . . .

. . . it was nevertheless unique and I wished to repeat it.

Having previously used spermatic vapor produced in closed vessels, I wished to see what would happen in open vessels in order to eliminate a doubt produced by the idea that the circulation of air was necessary to fertilization . . .

. . . but fertilization did not succeed any better than in the preceding experiments.

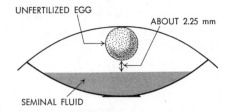

FIG. 3–5. Apparatus setup similar to the one used by Spallanzani to answer the question, "Is fertilization effected by the spermatic fluid?" Vapor rising from the seminal fluid freely bathed the egg, but no contact between egg and fluid occurred. The egg did not become fertilized.

The lack of development corresponds to a false conclusion; that is, the prediction which follows acceptance of the hypothesis being tested does not come true . . .

. . . and thus the vapor hypothesis must be false.

Spallanzani recognizes the need for further evidence that the vapor hypothesis is, indeed, incorrect. (His results in this second series of experiments were the same.)

A variable is recognized that might influence the results; the experiment is modified to eliminate it. If air plays a role in fertilization, then the eggs should develop if air is allowed to circulate, etc.

Again, negative results. The prediction is shown to be incorrect.

The last experiment of this type was to collect several grains of spermatic vapor and to immerse a dozen eggs in it for several minutes; I touched another dozen eggs with the small remnant of semen which remained after evaporation, and which did not weigh more than half a grain; eleven of these tadpoles hatched successfully although none of the twelve that had been plunged into the spermatic vapor survived.

The conjunction of these facts evidently proves that fertilization in the terrestrial toad is not produced by the spermatic vapor but rather by the material part of the semen.

As might be supposed, I did not do these experiments only on this toad, but I have repeated them in the manner described on the terrestrial toad with red eyes and dorsal tubercles, and also on the aquatic frog, and I have had the same results.   I can even add that although I have only performed a few of these experiments on the tree frog, I have noticed that they agree very well with all the others . . .

Shall we, however, say that this is the universal process of nature for all animals and for man?

The small number of facts which we have does not allow us, in good logic, to draw such a general conclusion.

One can at the most think that this is most probably so . . . ,

Still another variation of the original experiment is performed for further evidence against the vapor hypothesis; even immersion in a concentration of spermatic vapor does not result in fertilization.   Certainly the hypothesis being tested would have predicted fertilization in this case.   Yet, fertilization does not occur.

In their deductive format, Spallanzani's results do, indeed, show the vapor hypothesis to be false.   They do not, however, prove the validity of the alternative hypothesis, but only lend it support.   Note that Spallanzani is careful not to generalize beyond the animal used for his experiments.

Spallanzani now wishes to extend his results to other organisms, and so performs other experiments using different kinds of animals.

Spallanzani asks, ''Can the generalization be extended to other organisms not yet tested in these experiments?''

Spallanzani is cautious in considering an extension of his generalization regarding the necessity of contact with the semen rather than its vapor.

Spallanzani shows his awareness of the nature of scientific ''proof'' with this statement;

... more especially as there is not a single fact to the contrary ...

and the question of the influence of the spermatic vapor in fertilization is at least definitely decided in the negative for several species of animals, and with great probability for the others.*

that is, no wrong predictions (false conclusions) have been obtained in the experimental testing of this hypothesis.

Note the awareness on the part of Spallanzani that his negative results give him positive disproof of the hypothesis tested, yet give only probable verification rather than absolute proof to the opposite hypothesis.

Spallanzani later performed other experiments which further supported the results reported here. He discovered, for example, that if he filtered the semen through cotton, it lost much of its fertilizing powers, and that the finer he made the filter, the more the powers were diminished. He found, too, that several pieces of blotting paper completely removed the semen's ability to fertilize, but that the portion left on the paper, when put into water, *did* successfully fertilize eggs. Despite the obviousness (to us) of the role played by the spermatozoa in fertilization—a role to which the results of these experiments point—Spallanzani had previously decided that semen without spermatozoa *was* capable of fertilization, and he was unable to shake this belief ... even in the light of his own experimental results! If nothing else, this nicely demonstrates that scientists are just as prone to overlook the obvious solution as anyone else, and often refuse to give up preconceived notions despite evidence to the contrary. It was not until the nineteenth century that the role of the spermatozoa in fertilization was definitely established.

3-5   THE LIMITATIONS OF SCIENCE

In its own right, science is one of man's most productive ways of exploring, exploiting, and trying to understand his environment. But it is by no means the *only* way. The historian tries to understand the present and, occasionally, to predict the future, by studying the record of man's past. Religion attempts to find certain truths by operating mostly from a platform of faith. Philosophers draw on science, history, religion, and many other fields of endeavor in an attempt to consolidate the findings of each field and draw meaningful conclusions from them.

Further, we should note that despite the many contributions it has made to man's intellectual growth, as well as to his health and general welfare, science does have serious limitations. Oddly, one of these stems from one of its greatest

---

* From M. L. Gabriel and S. Fogel (eds.), *Great Experiments in Biology* (Englewood Cliffs, N.J.: Prentice-Hall, 1955). Used by permission.

attributes. As the philosopher George Boas points out,

> . . . *what science wants is a rational universe, by which I mean a universe in which the reason has supremacy over both our perceptions and our emotions.*

This rational basis of experimental science, rooted by necessity in concrete experiences, is, indeed, a strength. But it is also a weakness. By dealing only with that set of phenomena which can directly or indirectly be experienced through man's senses and placed into an experimental situation, science is necessarily excluded from that set of phenomena whose members do not have these qualifications. Experimental science can only attempt to explain *how* a natural phenomenon may occur, and hypothesize its causes. It cannot even begin to speculate *why* these phenomena occur, in the teleological sense of the word.

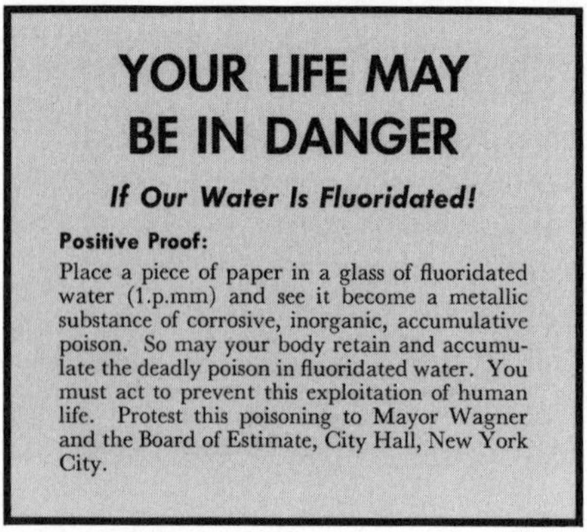

YOUR LIFE MAY BE IN DANGER

If Our Water Is Fluoridated!

Positive Proof:

Place a piece of paper in a glass of fluoridated water (1.p.mm) and see it become a metallic substance of corrosive, inorganic, accumulative poison. So may your body retain and accumulate the deadly poison in fluoridated water. You must act to prevent this exploitation of human life. Protest this poisoning to Mayor Wagner and the Board of Estimate, City Hall, New York City.

FIG. 3–6. As a field of endeavor, science is fairly immune to the blatant appeals to the emotions which an advertisement like this attempts to make. As an individual, however, the scientist is probably no more or less averse to emotional judgments than any other man.

The unemotional basis of science is another strength which is at the same time a weakness. This does not mean, of course, that scientists as individuals are at all unemotional or detached. They are in no way different from any other persons in this regard, being just as liable to personal prejudices and weaknesses as the next man. However, as a field of endeavor, science, at least in the long run, is necessarily objective and detached from emotional prejudice. Yet, there are times when man may not wish to let reason have supremacy over his emotions. Certainly we would not wish to deal with today's very real problems of human poverty and injustice in an entirely unemotional and detached manner. To retain its basic nature, however, and succeed in dealing with contemporary social problems, experimental science must do just that.

Despite the logical basis of science, it would be a mistake to give the impression that scientists are never wrong. Nothing could be further from the truth. The astronomer Johannes Kepler once wrote, "How many detours I had to make, along how many walls I had to grope in the darkness of my ignorance until I found the door which lets in the light of truth." It is doubtful, in fact, that there has ever been a scientist who has not made mistakes. James Bryant Conant states,

One could write a large volume on the erroneous experimental findings in physics, chemistry, and biochemistry which have found their way into print in the last one hundred years; and another whole volume would be required to record the abortive ideas, self-contradictory theories and generalizations recorded in the same period.

For example, Lord Rutherford stated that man would never tap the energy within the atomic nucleus. The first atomic bomb exploded a few years after his death. The famous nineteenth-century physiologist Johannes Müller asserted that the speed of the nerve impulse would never be measured. Six years later, Hermann von Helmholtz did it in a frog nerve only a few inches long.

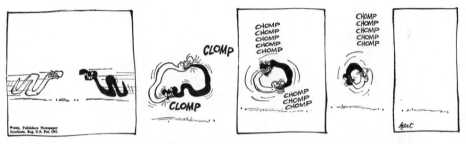

FIG. 3–7. The consequences of logical thinking do not always lead to "correct" conclusions. It is a well-known fact that snakes can and sometimes do swallow other snakes, even ones larger than themselves. If it is assumed that the one which is swallowed disappears from sight, the situation pictured above is the logical conclusion! (Courtesy Johnny Hart and Publishers Newspaper Syndicate.)

Nor is it safe to think that scientists always reason correctly. As a matter of fact, many scientists are notorious for "going off the deep end," particularly when writing in areas other than their own specialty. Furthermore, if scientists can be wrong, *so can science.* Ask a physicist about "ether," a chemist about "phlogiston," or a biologist about "lamarckism." Science has had incorrect theories in the past, it has them now, and it will continue to have them in the future. The strength of science does not lie in any infallibility. Nor does it lie in its logical basis, for the conclusion of a perfectly logical argument can be utter nonsense (see Fig. 3–7). Rather, it lies in the self-criticizing nature of science— the constant search for "truth" by the elimination of experimentally established error.

Science can be said to be a tradition of beliefs that have rational foundations, subject to continual review and discussion. Thus as a field, science is separate from the scientists who have contributed to its growth. As an individual, the scientist is only a human being, with all the emotions and weaknesses that are part and parcel of being so.

The scientist A. J. Lotka stated in 1925 that "Science does not explain any-thing . . . science is less pretentious. All that falls within its mission is to observe phenomena and to describe them and the relations between them." We might amplify this and state again that any absolute truth is beyond the reach of science. *In science, "truth" is a well-supported hypothesis.* Should the hypothesis fall, a new "truth" takes its place.

Yet, despite its built-in limitations, science must still be recorded as a remarkably successful way of accumulating knowledge. In **applied research,** scientists may use the methods of science for the purpose of developing products to improve human comfort and welfare. In **pure** or **basic research,** the scientist searches for knowledge—knowledge for its own sake—regardless of whether his discoveries will benefit mankind. It is important to note, however, that *the results of basic research have contributed as much as or more than those of applied research.* It seems that science is productive by its very nature.

3-6 CONCLUSION

Science can be considered as organized scientific knowledge or as the process by which such knowledge is obtained. Experimental science uses the format of deductive logic in the design of its experiments. The purpose of these experi-ments is to test the validity of the predictions which are made on the basis of accepting the hypotheses as true. If the experiment is properly designed, and the predictions are not borne out by the experimental data, the hypothesis may be considered disproved. If, on the other hand, the results of the experiment are as predicted, the hypothesis can be said to be supported, *but not proved.* The absolute proof attainable by the use of deduction in mathematics lies forever beyond the reach of experimental science. The validity of scientific "truths" can only be measured statistically in terms of probabilities.

The demand of science for rational, unemotional, experimentally demon-strable bases for its beliefs lend to its greatest strengths. At the same time, these demands limit science's working range to the perceivable and manipulatable environment. Science is only one of several highly successful ways available to man in his attempts to understand the universe and his role within it.

EXERCISES

1. Distinguish between basic (pure) and applied research. Why is it important for both kinds of research to be supported?

2. Explain why the attainment of any absolute truths lies beyond the realm of science.

3. List some of the limitations of science.

4. Why must a scientist be careful not to extend his experimental conclusions to organisms other than those with which he worked?

5. In a recent presidential campaign, some noted scientists and engineers formed a group to support one of the candidates. A spokesman for this group made the following statement:

   *By the time we were through, any guy in Pittsburgh in a T-shirt with a can of beer in his hand knew that the smartest people in this country considered —————— to be unfit [for the presidency].**

   What are the tacit assumptions made in this statement? Can you see any possible dangers which might develop from the acceptance of such reasoning?

6. Explain why the following hypothesis is an unacceptable one to scientists: Life originated on another planet somewhere else in the universe and came to earth millions of years ago enclosed in a meteorite.

7. Devise an hypothesis to explain each of the following observations. Then outline an experiment to test your hypothesis.

   a) There are more automobile accidents at dusk than at any other time.

   b) In washing glass tumblers in hot soapsuds and then immediately transferring them face downwards onto a cool, flat surface, bubbles at first appear on the outside of the rim, expanding outwards. In a few seconds, they reverse, go under the rim, and expand inside the glass tumbler.

   c) It has been noticed that one species of mud-dauber wasp will build its nests from highly radioactive mud, although the radiation received by the developing young may be enough to kill them. Another species of mud-dauber wasp, under the same environmental conditions, avoids this mud, and selects nonradioactive mud to build its nests.

   d) In mice of strain A, cancer develops in every animal living over 18 months. Mice of strain B do not develop cancer. However, if the young of each strain are switched immediately after birth, cancer does *not* develop in the switched strain-A animals, but *does* develop in the switched strain-B animals living over 18 months.

SUGGESTED READINGS

ARBER, AGNES, *The Mind and The Eye* (Cambridge: Cambridge University Press, 1964). A discussion of some facets of scientific work by a noted botanist. The book is one of the few attempts to discuss the philosophy of science using biological examples.

BRONOWSKI, J., "The Creative Process." *Scientific American*, September, 1958, p. 58. In this article the author tries to show that innovation in science is not different from innovation in fields of arts or social sciences. In all these cases, innovation comes from a person's seeing deep unity in a number of different situations.

CONANT, J. B., *On Understanding Science* (New Haven: Yale University Press, 1947). A clear, well-written account of the nature of science and its development. Uses many specific examples.

---

* *Science*, Vol. 146, No. 3650, p. 1444

CONANT, J. B., *Science and Common Sense* (New Haven: Yale University Press, 1961). This book is an attempt to make science more intelligible by showing the way in which a number of scientists came across their discoveries. Conant presents a number of "case histories" of scientific discovery.

GARDNER, M., *Fads and Fallacies in the Name of Science* (New York: Dover Publications, 1957). A well-written, very interesting account of a number of pseudoscientific theories, how they gained adherents, and how most of them can be shown to be invalid. Discusses "Bridey Murphy," "Atlantis," "flying saucers," Lysenkoism, and extrasensory perception (ESP).

HARDIN, GARRET, *Biology, Its Principles and Implications* (San Francisco: W. H. Freeman, 1961). An excellent discussion of scientific hypotheses, their modifications, etc., as well as many good examples are provided in this exceptionally well-written book.

STANDEN, ANTHONY, *Science Is a Sacred Cow* (New York: E. P. Dutton Co., 1950). A light-hearted debunking of the traditional view of science and scientists as infallible. Easy reading, written by a scientist.

TATON, RENE, *Reason and Chance in Scientific Discovery* (New York: John Wiley & Sons, 1962). A well-presented account of the role of chance, error, and inspiration in scientific discoveries, as well as an analysis of the relation between science and the culture of the times. A very intriguing and thought-provoking work.

The September, 1958 *Scientific American* is devoted to the topic "Innovation in Science." There are a number of articles in this issue on the growth of various sciences, and the conditions under which science is thought to flourish. The articles in this issue devote particular attention to the role of imagination and creativity in science (see also the Bronowski article above).

CHAPTER 4

TESTING HYPOTHESES AND PREDICTIONS

4-1 INTRODUCTION

Beri-beri is the common name for a degenerative and paralytic condition found in man and other vertebrates. Until about fifty years ago, this disease was relatively common in human populations, especially in the Far East and places such as Borneo and Java. Toward the end of the nineteenth century many people had voiced ideas as to what might cause the condition. The most popular idea was that beri-beri resulted from bacterial infection, for this was the period of the great influence of Louis Pasteur and his germ theory of disease. Bacteria were looked upon as the causal agent of all known human maladies.

In 1893 the Dutch government sent a commission to the East Indies to investigate beri-beri, which was particularly prevalent there. One member of that commission, Christian Eijkman (1858–1930), made observations which led him to formulate an alternative hypothesis as to the cause of the disease. Eijkman went on to design an experiment to test his hypothesis. This work provides a classic example of the design and execution of an experiment.

4-2 EIJKMAN'S EXPERIMENTS

Eijkman observed that experimental chickens kept around the laboratory were fed on a diet consisting mainly of polished rice. Many of these chickens seemed to have a condition which resembled beri-beri. Eijkman decided to see whether there might be any relationship between a diet of polished rice and the occurrence of beri-beri. He began by formulating two initial hypotheses:

Hypothesis I: Beri-beri is a result of dietary disturbance, and is not due to bacterial infection.

Hypothesis II: A factor present in rice husks seems to prevent appearance of the condition.

From these hypotheses Eijkman could make a simple prediction:

Hypothesis I: *If . . .*   beri-beri is a dietary condition, . . .
and

Hypothesis II: *If . . .*   beri-beri is a result of eating polished rice, . . .

Prediction: *then . . .* feeding chickens polished rice should produce the condition.   Conversely, feeding them unpolished rice should keep them healthy.

To test this prediction, Eijkman set up a simple experiment. He procured two groups of normal, healthy chickens. To one group he fed polished rice, to the other, unpolished rice. The chickens were placed in pens and kept under identical conditions for a period of two weeks. At the end of this period, many of those chickens fed on polished rice showed symptoms of beri-beri. Among those fed on unpolished rice, however, there were no symptoms. This experiment seemed to support Eijkman's hypothesis that beri-beri is a dietary condition.

Eijkman's experiment, simple as it is, has some very important features. First, it was designed to test a prediction made from a preliminary hypothesis. A good experiment is one designed to test a specific prediction, which the experiment can then either confirm or reject. If the prediction from a given hypothesis is verified, the hypothesis may be correct—although, as we saw in Chapter 3, it is not *necessarily* so. On the other hand, if the prediction proves to be false, then the hypothesis itself *must* be false. Note that the most conclusive experiments are often ones which disprove a specific prediction, for they make it possible to reject the hypothesis from which the prediction was drawn.

Second, Eijkman's experiment made use of *controls.* In a controlled experiment, two groups of organisms are treated alike in all but one respect. The one difference is the factor being studied, such as polished or unpolished rice in the diet. In Eijkman's experiment, the chickens fed on polished rice form the **control group**, while those fed on unpolished rice form the **experimental group**. In general, control groups represent the normal situation, whereas experimental groups represent the *variation*. A control group provides a basis for comparison—a standard against which changes in the experimental group can be measured.

Scientific experiments are generally designed to test one prediction at a time. If Eijkman had set up his experiment as described above, but had used ducks for his experimental group and chickens for control, this would have introduced *two variables* into the experiment: the difference in diet and the difference in type of organism. If the experimental group had shown a high incidence of beri-beri, Eijkman would have been much less certain of his results than in the experiment as actually performed. It could be argued, for instance, that ducks are highly susceptible to beri-beri, and contract the condition under circumstances where chickens do not. This is not a far-fetched objection; many diseases are known to appear frequently in some organisms, and seldom or never in others. Thus, if he had used two different types of organism, Eijkman would not have been able to conclude that beri-beri was a dietary condition.

Although Eijkman's original experiment did support his hypothesis, it did not eliminate all other possibilities. Eating polished rice, for example, might merely lower an animal's resistance to infectious organisms. Thus chickens fed on polished rice would be more susceptible to beri-beri germs, but would contract the condition only if such infectious organisms were present. Although subsequent research showed that this possibility was not correct, the experiment described above did not provide evidence to that effect. Furthermore, Eijkman could not conclude from his experiment that because a diet of polished rice caused beri-beri in chickens, it must necessarily have the same effect on other animals. He needed to test his hypothesis in relation to human beings.

It is normally difficult to carry out large-scale experiments with human beings, especially under conditions which are likely to produce harmful effects. An already existing situation, however, helped Eijkman solve this problem. Beri-beri was especially common among the penal institutions of Java. Eijkman ordered unpolished rice to be introduced into the diets of the inmates in a few of those prisons where beri-beri was quite common. A definite improvement was shown, but still the results were not conclusive. There was always the possibility, however remote, that the prisoners would have recovered anyway, due to changes in the nonrice portion of their diet during the period under observation. A control group was needed.

Luckily, local customs provided just what was needed. Prison diets throughout Java were remarkably similar, except for one factor. Although rice was a staple in every prison, in some areas polished rather than unpolished rice was used. By collecting information gathered by the Supervisor of the Civil Health Department of Java, Eijkman had data for his control and experimental groups right at hand. Data were available for 100 prisons throughout Java and a neighboring island. It was important to have as many experimental situations as possible. As we saw in Chapter 3, the more data there are from which to work, the more valid a generalization will be. Data from 100 prisons are adequate to avoid any significant source of error.

From this sample Eijkman found the following results: Out of 27 prisons where unpolished rice was fed to prisoners, beri-beri occurred in only *one* prison. On the other hand, beri-beri occurred in 36 out of the remaining 73 prisons, where polished rice was a main part of the diet. Then, finding that actually three rice diets were being used, Eijkman made a new classification of diets:

1) polished rice (husks entirely or at least 75% removed),

2) unpolished rice (husks entirely or at least 75% preserved),

3) a mixture of (1) and (2), being served in some prisons.

TABLE 4-1

DIET CONSISTING OF	PERSONS CONTRACTING BERI-BERI, %
POLISHED RICE	70.6
UNPOLISHED RICE	2.7
MIXTURE OF POLISHED AND UNPOLISHED RICE	46.1

Data for this new series of categories were tabulated as shown in Table 4–1. It is apparent that Eijkman's hypothesis (that beri-beri is caused by a dietary factor) is well supported by these data.

However, these experiments did not answer all the questions which this investigation into beri-beri had raised. For example, it was suggested by some people that beri-beri was caused by eating old spoiled rice; others said that rice imported from Saigon or Rangoon was the causative agent. Still others claimed that hygienic factors in the various prisons were the cause. Eijkman was able to answer all these objections, and thus gained even more support for his hypothesis. He fed chickens polished and unpolished rice of all sorts—old and new, from Saigon and from the East Indies. In all cases, those chickens which showed beri-beri were fed on polished rice.

In meeting other objections, Eijkman studied the age of prison buildings, their ventilation, and the population density in all prisons. His findings for age of building and ventilation are given in Table 4–2. It can be seen that there is no relation between either of these factors and the percentage of persons contracting beri-beri. These data effectively ruled out hygienic factors as possible causes of beri-beri.

In addition to those features discussed earlier, Eijkman's experiments illustrate two additional features of some importance in experimental analysis. Eijkman used a large number of samples in order to avoid *sampling error*, and he collected *numerical data*. The importance of using a large number of samples has already been discussed. The importance of numerical data deserves special mention.

Information collected from observation or experiment may be of two kinds: *quantitative* and *qualitative*. Quantitative data are those which are the result of measurement, or which can be expressed in some definite and precise form, usually in numbers. Qualitative data, on the other hand, are those data which do not lend themselves to precise numerical expression, Differences in height of a group of people, stated in terms such as "taller," "shorter," "tallest," etc., would be an example of qualitative data. The same differences, stated as results of measurement of height (as in inches or centimeters), would be an example of quantitative data.

TABLE 4–2

AGE OF BUILDINGS	PERSONS CONTRACTING BERI-BERI, %
40–100 YEARS	50
21–40 YEARS	34.4
2–20 YEARS	45.2

VENTILATION	
GOOD	41.2
MEDIUM	72.7
FAULTY	33.3

In general, scientists prefer numerical data because they are more easily subject to verification. Suppose that an experiment is performed in which the amount of growth of plants is studied under different intensities of light, and that the results are presented as follows: Plant #1 grew a little under a dull light, Plant #2 grew somewhat more than this under a medium-bright light, and Plant #3 grew very much under a bright light. These findings are based on subjective judgments. It would be difficult to repeat the experiment and know whether the new results agreed with the old. On a numerical basis, the data might be presented as follows: Plant #1 grew 5 inches under light intensity of 500 candlepower; Plant #2 grew 8 inches under light intensity of 700 candlepower; Plant #3 grew 10 inches under light intensity of 1000 candlepower, etc. This kind of data is easy to check. The experiment can be performed again, and the degree of difference or similarity between the two sets of data can be observed.

Numerical data are also valuable because relationships between two factors are more readily apparent (as between light intensity and amount of growth). In the above case, the fact that plants show greater growth under high light intensity than under low light intensity is more evident when the data are presented in numerical form. Although subjective data do not necessarily obscure such a relationship, they make it much less apparent.

Finally, numerical data are valuable because they make possible more precise and meaningful communication between scientists. To say that a plant grows "a little" or "very much" may mean different things to different people. Hence ambiguity arises. The use of numerical data prevents such ambiguity, and the results of experiments are less easily subject to misunderstanding.

4-3 CHANCE AND TRIAL-AND-ERROR IN SCIENTIFIC DISCOVERY

In this chapter and the preceding one, we have stressed the logical, planned approach to scientific discovery. This is indeed a very important part of scientific research. Yet, other factors are involved in science which cannot be called logical or predictable. The roles played by trial-and-error and chance in scientific discovery have often been neglected in discussions of scientific procedure, but both these factors play important roles.

The role of trial-and-error. A rat in a maze tries first one passage, then another, until it reaches the end and is rewarded by a piece of cheese. If the rat has no previous experience with the maze, the only method it can use to reach the cheese is that of trial-and-error. This method is one which is frequently thrust on an organism, be it rat or man, that is faced with solving a problem for which no definite and clear-cut clues are available.

Consider the following example of trial-and-error in scientific research. Dr. Paul Muller, who won the Nobel Prize in 1948 for his discovery of dichloro-diphenyl-trichloro-ethane (DDT), said in his acceptance speech: "After fruitless testing of hundreds of different substances, I realized that it is not easy to find a good contact insecticide." The problem was to find a substance which would be lethal to insects but relatively harmless to man and animals. For Muller, there was no other way than to test many substances, one after the other, until he finally hit upon DDT, which seemed to be effective. The discovery was made only by long hours of testing a variety of possible compounds.

The method of trial-and-error has one disadvantage. It is inefficient. If an experimenter has some clues from which he can formulate an hypothesis at the beginning, he can reduce the number of experiments which he has to perform. In fact, if all knowledge were gained only by trial-and-error, scientific advances would be very slow in coming.

The role of chance. To ignore the role of chance discovery would be to present a false picture of scientific research. There is a difference, however, between being lucky and being observant enough to take advantage of this luck. The history of science is filled with examples of "missed discoveries." Although many individuals have hit upon important discoveries during the course of their investigations, they ignored their discovery because they were looking for something else. To some extent, an investigator must be prepared to seize upon chance observation.

An interesting case of the role of chance occurs in the work of the French biologist Louis Pasteur (1822–1895) on the problem of immunity. In 1798, Edward Jenner had shown that animals and people could be immunized against the infectious disease smallpox by injection of small amounts of material extracted from the sores of cattle affected by a similar condition known as cowpox. This was a specific case, however, and Jenner had not attempted to extend his findings to other diseases. However, by the middle of the nineteenth century the question of a more general application of Jenner's technique to immunization against all sorts of diseases was widely discussed in medical circles. Could immunization be established for other diseases caused by microorganisms, just as Jenner had shown it could be done for cowpox? Pasteur had pondered this problem for several years, while engaged in other fields of research. Then a chance occurrence gave him the clue from which there developed a whole theoretical science of **immunology**.

Pasteur had begun experiments on chicken cholera in the spring of 1879, but had discontinued his work during the summer. When he returned toward the beginning of September, the cultures of chicken cholera bacteria (which had been kept in the laboratory untended for several months), failed to produce the disease when injected into chickens. A new, virulent (capable of causing disease) culture was obtained and used to inoculate not only new animals, but also the chickens which had been inoculated previously with the old culture. The new animals quickly contracted the disease as expected. To his surprise, however, Pasteur found that the previously inoculated chickens showed no signs of cholera and remained quite healthy.

As a result of his own reading and thinking on the general problem of immunity, Pasteur immediately recognized the similarity between this situation and Jenner's immunization of animals against smallpox. The evidence was as yet too scanty for any really valid conclusions to be drawn. But Pasteur was sufficiently thoughtful to suspect the general application of a new principle. By transferring pox material taken from an infected cow to man, Jenner had altered the human constitution so as to render it "immune" to infection from other pox microorganisms. Pasteur now recognized that the Jenner effect was a manifestation of a general law. The old bacteria culture which had remained in Pasteur's laboratory over the summer had lost its ability to produce the disease symptoms in an organism. But it had not lost the ability to elicit from the host animal the immunization response which makes that organism unreceptive to virulent microorganisms of the same type in the future. Jenner had used the term "vaccine" to refer to the substance removed from the sores of diseased cows for injection into other animals and people. Pasteur now coined the term "vaccination" to refer to the creation of immunity by similar means in any organism against any of a number of infectious diseases.

Pasteur's chance observation allowed him to make a speculative leap from Jenner's specific case to the general biological principle that nonliving microorganisms of infectious diseases can be utilized to build an immunological response in host animals. Chance was not all that was involved, however; Pasteur's own awareness of the *importance* of his observation made possible his recognition of such a generalization. In other words, chance is often a key ingredient in scientific discoveries, but it is seldom enough by itself to produce an important theory or idea. As Pasteur himself pointed out, "Chance favors the prepared mind."

4-4 A CASE HISTORY OF SCIENTIFIC EXPERIMENTATION

Having gained some general ideas as to the character and properties of scientific experimentation, we may now proceed to analyze a complete research project. This will provide a better understanding of the way in which observation, experiment, and logical reasoning interact to lead to a final conclusion. As an example, we shall consider the work of two English physiologists, W. M. Bayliss and E. H. Starling. Their experimentation was simple but ingenious, and thus serves as an excellent model to illustrate some of the important characteristics of scientific work. However, in order to understand exactly what is involved in the work of Bayliss and Starling, it is first necessary to be familiar with (1) the general anatomy of the digestive system in the area of the pancreas, and (2) the nature of the dispute among physiologists which led to the formulation of the basic problem in the minds of Bayliss and Starling.

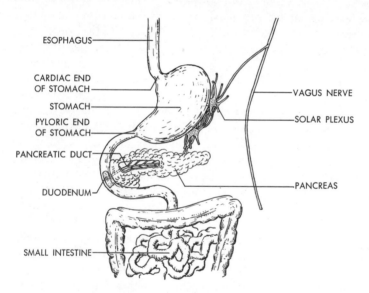

FIG. 4–1. Generalized scheme of the alimentary tract from the esophagus to the upper part of the small intestine. Shown also is the vagus nerve on the left-hand side of the body, and the solar plexus, a reflex center for many organs in this region of the abdomen.

In the early part of this century, Bayliss and Starling had discovered that the pancreas, one of the digestive organs, is stimulated to release its digestive enzymes at precisely the right moment, as food passes from the stomach into the upper portion of the small intestine. The question left unanswered was basically, "What is the mechanism by which the pancreas is stimulated to release its digestive juice at precisely this moment?"

In the normal digestive process, food passes from the mouth into the **esophagus**, a long tube which leads to the stomach. Figure 4–1 shows the basic organization of the alimentary canal in the region of the pancreas. At the point where the esophagus joins the stomach (the **cardiac** end of the stomach), there is a special type of muscle which, when contracted, acts like a drawstring, closing off the stomach from the esophagus. Another valve, at the lower or **pyloric** end of the stomach, serves a similar function in closing the stomach off from the intestines. Inside the sealed-off stomach, the food is churned by the action of muscles of the stomach itself and mixed with gastric juices secreted from the tissue which lines the stomach walls. When the food is thoroughly mixed and partly broken down, the pyloric valve opens. The food passes into the **duodenum**, a portion of the upper end of the small intestine approximately 10 to 12 inches long. At this point, the food is in a semiliquid condition known as **chyme**. Opening into the duodenum is a duct leading from the pancreas. The pancreas produces a digestive juice which contains enzymes specialized to act upon each of the major types of food: carbohydrates, fats, and proteins. Within several minutes after the chyme has entered the duodenum, the pancreas begins to secrete its digestive juice. This juice mixes with the chyme, and the entire mass is moved slowly through the small intestine.

The following information was available at the time Bayliss and Starling began their work. If the pyloric valve was tied off so that food could not pass from the stomach into the duodenum, there was no pancreatic secretion. Thus the passage of food from the stomach into the duodenum somehow provided the signal to begin pancreatic secretion. It was also known that the chyme passing into the duodenum was highly acidic. It was suggested that the acid nature of the chyme was responsible for triggering pancreatic secretion. Previous workers had shown that introduction of dilute hydrochloric acid into the duodenum of an anesthetized animal brought about prominent activity of the pancreas. Introduction of similar but nonacid control substances yielded no secretory activity.

Given this information, two hypotheses were put forward in the late nineteenth century to explain the exact mechanism by which the chyme, as it passed into the duodenum, caused the pancreas to begin secretion.

Hypothesis I: The pancreas is controlled by the nervous system. Food entering the duodenum stimulates nerve endings in the walls of this organ. These nerves carry an impulse to various centers in the spinal cord and brain. From these central relay points, appropriate stimuli are sent back to the pancreas, bringing about a release of digestive juice. The basic nerve-reflex pattern is shown in Fig. 4-2.

Hypothesis II: The stimulation is carried from the walls of the duodenum to the pancreas by means of a "chemical messenger" in the blood. This messenger is perhaps produced by the cells in the wall of the duodenum.

Evidence for hypothesis I came from the work of several earlier physiologists. They had shown that stimulation of certain nerves, especially those in a prominent tract known as the *vagus*, caused an increase in pancreatic secretion. It was also known that the vagus did not stop pancreatic secretion.

Those adhering to hypothesis II also had good evidence to support their view. One of the most interesting experiments was performed in Germany in the 1870's. Two dogs were anesthetized and joined together in such a way that their blood streams were interconnected. When the duodenum of *one* of these two dogs was exposed and injected with dilute acid, the pancreas of both dogs began to secrete pancreatic juice. Since there were no nervous-system interconnections between the two animals, the experimental result was good indication that a blood-borne messenger was indeed responsible for triggering the pancreas.

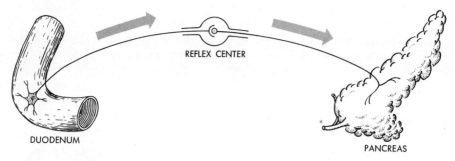

FIG. 4–2. Schematic representation of a reflex system, specifically referring to that idea hypothesized before the work of Bayliss and Starling. A nerve leading from the duodenum passes to a ganglion (reflex center), such as the solar plexus. From here, the message is relayed out to appropriate nerve tracts leading to other organs, in this case to the pancreas. The reflex center thus acts like a telephone switchboard.

With two opposing hypotheses to explain the same phenomenon, there was need for a crucial experiment, i.e., one that would serve to disprove one hypothesis. Recall from Chapter 3 that a crucial experiment is one based on a point of disagreement between two rival hypotheses. If two such hypotheses are compared point for point and a situation found where each hypothesis yields a different prediction, a crucial experiment may be designed around this point of difference. Bayliss and Starling developed such a test. They reasoned:

Hypothesis: If . . .   the pancreas is stimulated by a chemical substance in the blood rather than by central or peripheral nervous reflexes, . . .

Prediction: then . . . removal of all ganglia and severing of all possible nervous pathways between the duodenum and the pancreas should not interfere with pancreatic secretion when acid is introduced into the duodenum.

This was the point of difference between the two hypotheses, since according to hypothesis I, if all possible nerve pathways were cut from the duodenum, then injection of acid should fail to elicit the normal pancreatic secretion.

In 1902, Bayliss and Starling performed their crucial experiment. Opening the abdomen of an anesthetized animal, they removed all the nerves from the duodenum and cut the vagus nerves on each side of the body. They carefully tied a loop of the duodenum in two places, so that it was no longer connected to other parts of the digestive tract by any direct opening. All other nerve connections to the duodenum were also carefully cut away; thus this particular loop of the duodenum was connected to the body of the animal only by its arteries and veins. A glass tube, its free end leading out of the animal's body, was inserted into the pancreatic duct so that the number of drops of pancreatic juice secreted for any unit of time could be accurately recorded on a revolving drum or **kymograph**. The experimental setup is diagrammed in Fig. 4–3.

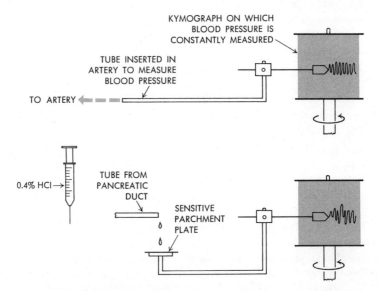

FIG. 4–3. Experimental setup for the Bayliss and Starling experiment. Injection of acid solution into the duodenum produces flow of pancreatic fluid, measured on drum at lower right. All connections to the duodenum from the rest of the body have been severed except arteries and veins.

To ensure against possible sources of error, Bayliss and Starling took the precaution of recording, on a separate revolving drum, the blood pressure of the animal under investigation. It is important to see why this step was necessary. For their normal functioning, tissues depend on a certain amount of blood flowing through them at a certain pressure. If the pressure falls, materials in the blood do not move from the capillaries to the tissues as rapidly, and thus the physiological activity of these tissues is altered. If the blood pressure in the experimental animal were to fall during the course of the experiment, the amount of pancreatic juice would also be expected to fall. By keeping a running record of the dog's blood pressure, the experimenters would be able to tell when the reduction of pancreatic juice was a result of a change in blood pressure, and when it was due to some condition of the experiment.

Injecting a small amount of dilute (0.4%) hydrochloric acid into the isolated loop of the duodenum, Bayliss and Starling found that after a delay of two minutes or so, the pancreas began to secrete digestive juice at a relatively rapid rate. It seemed apparent that the stimulus must be carried to the pancreas through the blood rather than by way of a nervous pathway.

As a control, Bayliss and Starling recorded the rate of pancreatic secretion in an anesthetized animal whose nervous connections to the duodenum and pancreas were left intact. This rate, brought about by the introduction of 0.4% HCl into the duodenum, served as a comparison for the rate observed in the experimental animals. Note that the chief difference between the experimental and control animals was the absence or presence of nervous connection between the duodenum and pancreas. Thus a similarity in the rates of secretion in the two animals would indicate that nerve impulses were not the major factor in bringing about the activation of the pancreas, because *if* nerve pathways were involved in stimulating the pancreas, *then* the rate of flow would be greater in the control (nerves intact) than in the experimental (nerves severed) animal. Since the rate of flow in both turned out to be the same, Bayliss and Starling concluded that it was possible to account for the stimulation of the pancreas by mechanisms other than the nervous reflex. Hypothesis I was thus disproved, and hypothesis II supported.

But this was not enough. Once the nerve-pathway hypothesis was ruled out, it became an important phase of Bayliss and Starling's work to show exactly what the hypothetical blood-borne messenger was, where it was produced, and how it had its effects. These problems formed the foundation for the next phase of experimentation.

It was known that direct injection of hydrochloric acid into the blood did not increase the rate of pancreatic secretion. Hence the acid itself was not the immediate stimulus. Bayliss and Starling now concentrated on the tissue of the wall of the duodenum and intestine. Since the acid chyme entered the duodenum from the stomach and came into contact with the epithelial (lining) cells which composed the wall of the duodenum, it was reasonable to assume that perhaps the chyme stimulated the production of some chemical substance in the cells of the intestinal walls. This chemical messenger, then, carried in the blood to the pancreas, served as the stimulant to trigger the release of digestive juices by the latter organ.

The experimenters tested this hypothesis in the following way. They scraped some of the epithelial cells from the inside wall of the duodenum and intestine, and mixed these scrapings with dilute hydrochloric acid. Then they filtered this liquid and injected it into the bloodstream. A minute or two thereafter, an enormously large secretion of pancreatic fluid was recorded, indicating that the action of acid on the epithelial cells caused the formation of a messenger substance which passed into the blood. Although this substance passed to all tissues of the body, it would stimulate only the cells of the pancreas itself. Bayliss and Starling called the messenger substance **secretin**. Today we know that secretin is one of many substances, called **hormones**, which act in this general way. A hormone is a chemical messenger produced by one area of the body (usually by an **endocrine gland**) to have some influence in another, very specific area. Certain cells of the walls of the duodenum thus constituted an endocrine gland producing the hormone secretin.

Further investigation showed that if the scrapings were taken from successively lower portions of the intestine and mixed with acid, the magnitude of their effect on the pancreas decreased. Thus the lower ends of the small intestine would not produce any secretion when brought in contact with dilute hydrochloric acid. In order to explain exactly how this might occur, Bayliss and Starling hypothesized that the cells of the upper portions of the intestinal tract contained a large amount of a substance they called *prosecretin*, which was converted into secretin by the action of dilute acid on the cells. They expressed this as:

PROSECRETIN + HYDROCHLORIC ACID → SECRETIN.

The further along the intestine the cells were located, however, the less prosecretin they contained, hence the less secretin they produced when stimulated by acid. Finally, the cells of the lower end of the intestine contained no prosecretin whatsoever, and hence could not function as a hormone-producing gland. It was logical that the most active area of the intestine should be the duodenum—that area closest to the lower end of the stomach. There the acid food has its greatest effect, and thus causes the relatively rapid secretion of digestive juice by the pancreas.

The work of Bayliss and Starling served to end a long-standing controversy. For this reason it is important in the history of physiology. In addition, it highlights many of the features which often are part of scientific experimentation. Let us consider some of the main features which this experimental work emphasizes.

1) Bayliss and Starling were led to their work because of an existing controversy. This controversy developed between two rival hypotheses, both of which might explain the phenomenon of pancreatic secretion in an acceptable way. Comparing the two hypotheses point for point, Bayliss and Starling were able to find an area in which one theory predicted one result, while the other predicted something different. They were thus able to organize a crucial experiment which served to settle the controversy. In other words, they formulated their reasoning in such a way that only *one question* was at stake in the experimental design: Would the pancreas secrete digestive juice if the duodenum were isolated from the animal in all ways except through the bloodstream?

2) The experimenters were able to carry the investigation further than answering just a single question. They were able to suggest a physiological mechanism by which the acid chyme from the stomach actually produced the stimulation of the pancreas by way of the bloodstream. This enhanced their original finding in the crucial experiment because it offered a replacement for the nerve-pathway hypothesis—some type of chemical explanation which was consistent with their

new findings. This is an important feature of scientific experimentation. It is not enough to destroy an old hypothesis; it is also necessary to carry the work far enough to offer some type of replacement theory which is in agreement with all available data. In other words, a crucial experiment should not only explain all the old findings, but also any new findings which the old hypothesis could not explain.

3) Bayliss and Starling set up their experiment in such a way as to record their data precisely and quantitatively. By using the parchment plate and the revolving drum, they were able to get an accurate record of the number of drops of digestive juice which the pancreas secreted under different experimental conditions. They also took the precaution of recording the blood pressure, so that any changes which this might make in the activity of the pancreas could be accounted for. They tried to leave no loopholes which could detract from the significance of their findings.

4) The experimenters introduced a control factor, the animal in which the nervous connections between duodenum and pancreas were left intact. In this way the rate of pancreatic flow with the nerves severed could be compared to some standard. The control was important, in this case, since a major purpose of the work of Bayliss and Starling was to show that nerve reflex was not a significant cause of pancreatic secretion.*

4-5 CONCLUSION

In this chapter, we have gained some insight into the nature of scientific experiments, their characteristics and their design. The characteristics discussed are not the only ones found in good experiments, nor do all valid experiments *necessarily* have these features. Yet, all good experiments do have certain things in common. The validity of the experimental results rests upon the degree to which the important criteria of experimental design are met.

EXERCISES

1. What is the relation between an hypothesis and an experiment?
2. What is a controlled experiment? Why is it essential to have controls whenever possible in experimental work? List some of the most important characteristics of a controlled scientific experiment.

---

* For a thorough analysis of the Bayliss and Starling work, it is suggested that the student read an abbreviated version of the original paper of 1902, listed in the readings at the end of this chapter. This paper, incidentally, can serve the reader as a model for the general format of scientific articles.

3. A biologist finds that removal of organ A, an endocrine gland, from an adult mammal causes organs B and C to cease functioning. Organ B is also an endocrine gland. The three possible explanations for this occurrence are diagrammed below. (For A → B, read "A is necessary to B," etc.)

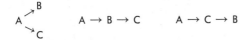

Design an experiment or experiments which would test these possibilities and which would distinguish between them.

4. In order to give his experimental results validity, a scientist wanted to test the effectiveness of a certain vaccine. He went to a village which was made up of equal numbers of natives and persons of another race. The vaccine was supposed to prevent a certain disease to which the entire village population was susceptible. Which of the following should he do in order to test his serum in a valid way?

   a) Give vaccine to the natives but not to the non-natives, and watch the results.

   b) Give vaccine to the non-natives but not to the natives, and watch the results.

   c) Give vaccine to the natives and a harmless salt solution to the non-natives, and watch the results.

   d) Give half of the natives and half of the non-natives the serum, and half of the natives and half of the non-natives a harmless salt solution, and watch the results.

   e) You cannot run a validly controlled experiment with human beings, since they are so complex.

5. From the following experimentally obtained observations regarding mineral nutrition in plants, draw a conclusion as to the factor or factors necessary for the development of chlorophyll in green plants.

   Observation 1. Plants grown in soil containing chloride and magnesium, and supplied with light, became green.

   Observation 2. Plants grown in soil containing chloride but not magnesium, and supplied with light, remained white.

   Observation 3. Plants grown in soil containing chloride and magnesium, but kept in the dark, remained white.

   Observation 4. Plants grown in soil containing magnesium but no chloride, and supplied with light, became green.

   Observation 5. Plants grown in soil containing chloride but not magnesium, and kept in the dark, remained white.

   Observation 6. Plants grown in soil containing neither magnesium nor chloride, but supplied with light, remained white.

   Observation 7. Plants grown in soil containing magnesium but not chloride, and kept in the dark, remained white.

   Observation 8. Plants grown in soil containing neither chloride nor magnesium, and kept in the dark, remained white.

Conclusion: The factor (or factors) necessary for the development of chlorophyll, as judged from the above experiment, are: . . .

6. A few years ago a number of psychological experiments were performed with a "consciousness-expanding drug" known as psilocybin. This drug, an extract from certain fungi, heightens the senses, makes a person more aware of his surroundings, and supposedly gives each individual a deeper understanding of himself. To measure the effects of the drug, several people, including the experimenter himself, would take a dose simultaneously. Researchers claimed that only when the experimenter was also under the influence of the drug could he properly evaluate the subjects' responses. Discuss the way or ways in which such an experiment fails to meet the requirements of a valid scientific experiment.

7. A group of scientists wanted to know what factor or factors cause rats to be susceptible to a certain disease induced by viruses.

Hypothesis: Diet is the factor which is responsible for susceptibility to the condition induced by a particular virus.

Experimental Procedure: Six pens were prepared, each to house 20 laboratory rats chosen at random from a stockroom. A special food was prepared, containing an adequate amount of carbohydrates, fats, proteins, vitamins, and minerals. The rats in the six pens were subjected to the following feeding formulas:

Pen A – These 20 rats were fed the special food (F).

Pen B – These 20 rats were fed the special food (F) with carbohydrates omitted.

Pen C – These 20 rats were fed the special food (F) with both fats and carbohydrates omitted.

Pen D – These 20 rats were given food containing vitamins and minerals only.

Pen E – These 20 rats were given food containing minerals only.

Pen F – These 20 rats were given no food at all.

Evaluate each of the occurrences or results I–IV on the basis of such considerations as the following:

    a) Is it a logical step in experimental procedure?

    b) Is there adequate control of all variables, or are additional variables introduced?

    c) Is the result observed (if it is a result) expected or unexpected?

    d) Is the occurrence related to the success or failure of the experiment?

I. The rats used to introduce the virus into the rats in pens A, B, C, D, and E were trapped at the city dump and placed in the pens as soon as they showed symptoms of virus V.

II. The rats in pens E and F appeared to lose weight most rapidly but failed to show symptoms of virus V.

III. All the rats in pen F outlived all the rats in all the other pens.

IV. Adequate water and suitable temperature were provided for all the rats in the six pens throughout the experiment.

SUGGESTED READINGS

BAYLISS, W. M., and E. H. STARLING, "The Mechanism of Pancreatic Secretion." *Journal of Physiology* 28:325–353. Reprinted in *Great Experiments in Biology*, M. L. Gabriel and S. Fogel, eds. (Englewood Cliffs, N.J.: Prentice-Hall, 1955), pp. 60–64. This is a reprint of Bayliss and Starling's original paper, somewhat abbreviated. A good insight into how they designed the crucial experiment.

DUBOS, RENE, *Louis Pasteur, Free Lance of Science* (Boston: Little, Brown & Co., 1950). This biography and the shorter paperback version, *Pasteur and Modern Science* (New York: Anchor Books, 1960), provide further insight into Pasteur's many areas of research, and his experimental approach to problems. Both books are very well written, and show the roles of chance and luck, coupled with Pasteur's own genius, in leading him to so many important discoveries.

GOLDSTEIN, PHILIP, *How to Do an Experiment* (New York: Harcourt, Brace, 1957).

YOUDEN, W. J., *Experimentation and Measurement* (Washington: NSTA Books, 1962). A valuable and well-organized booklet containing much information about the design of experiments and the handling of biological data.

CHAPTER 5

THE ANALYSIS AND INTERPRETATION OF DATA

5-1 INTRODUCTION

As important to the scientist as the collection of data is how he extracts from it the significant information it may contain. A large collection of data is of limited value if it is not arranged in such a way as to show important relationships. For example, random samples of age and height measurements for a population of human beings would not show what relation exists between these factors. Grouping the measurements by age in one column with corresponding height in another would yield much more information. It would show, for example, the ages during which growth occurred most rapidly, as well as the ages at which growth slowed down and ceased.

As a means of extracting as much information as possible from their measurements, working scientists subject their data to many kinds of analysis. They may arrange their data in tables or graphs, or subject them to specific mathematical procedures known as statistical analysis. Such treatment provides the researcher with a more complete understanding of his experimental results and of the scientific principles involved. This chapter will discuss some of the ways in which data can be handled, and how such procedures are of great importance in scientific work.

5-2 FRAMES OF REFERENCE FOR BIOLOGICAL DATA

Measurements are the bases of a quantitative approach to biology. Any system of measurement establishes a basic frame of reference within which comparisons can be made. In the United States, for example, the common units of linear measurement are the inch, the foot, the yard, and the mile. Within this frame of reference, comparisons of distance can be made with relative ease because these units of measurement have become familiar to everyone.

A single unit for measuring distance will not usually be convenient for all possible situations. Different units must be employed in making different kinds of measurements. For example, the astronomer uses the *light-year* as his basic unit for measuring distances outside our solar system. A light-year is the distance which light can travel in one year's time. The distance to the nearest star (Alpha Centauri) can be expressed conveniently as 4.4 light-years, whereas if it were expressed in terms of miles, the figure would be very awkward to work with.* Similarly, scientists in other fields use different units of measure for distance, volume, or mass, the choice depending on the magnitudes which they wish to measure. One unit can generally be converted into another, if necessary. For example, a light-year *can* be expressed in terms of miles if this is desired. The most convenient frames of reference are those where one unit can be converted into another with relative ease.

Nearly all scientists use the **metric system** of measurement. In the metric system, the standard unit of linear measure is the **meter,** equivalent to 39.37 inches. The standard unit of mass is the **gram,** the mass of one milliliter of water at a temperature of 4° centigrade. The standard unit of volume is the **liter,** equivalent to approximately 1.06 quarts.

The metric system has two important advantages. First, all units are divided and subdivided on the basis of tens. Thus 1 liter is composed of 10 deciliters, or 1000 milliliters. Similarly, 1 meter is composed of 10 decimeters, 100 centimeters, or 1000 millimeters, and 1 gram is composed of 10 decigrams, 100 centigrams, or 1000 milligrams.

A second advantage of this system is that units can be easily converted from one type of measure to the other; thus the capacity measure 1000 milliliters (ml) equals the volume measure 1000.027 cubic centimeters (cc or cm^3), and so on. For ordinary purposes, 1000 ml is considered equivalent to 1000 cc. Such flexibility greatly simplifies routine calculations.

One of the very helpful features of the metric system is that the unit names are to some extent *descriptive* of the unit values. For example, a centimeter is one-hundredth of a meter (from the Latin word *centum,* meaning "hundred"), and a millimeter is one-thousandth of a meter (from the Latin word *mille,* meaning "thousand"). The prefix *micro-,* which literally means "small," has come to mean one-millionth; thus a microgram is one-millionth of a gram. The Greek prefix *kilo-,* meaning "thousand," is used in the multiple rather than the fractional sense. Thus a kilometer is 1000 meters, and a kilogram, 1000 grams. Table 5–1 compares some of the major metric units. English equivalents for most of these metric units are shown in the right-hand column of the table.

---

* Light travels about 5,865,696,000,000 miles per year. To say that the nearest star is 25,809,062,400,000 miles away is so cumbersome as to be almost meaningless.

TABLE 5–1 CHIEF UNITS OF THE METRIC SYSTEM

LINEAR MEASURE		SYMBOL	ENGLISH EQUIVALENT
1 KILOMETER	= 1000 METERS	km	0.62137 MILE
1 METER	= 100 DECIMETERS	m	39.37 INCHES
1 DECIMETER	= 10 CENTIMETERS	dm	3.937 INCHES
1 CENTIMETER	= 10 MILLIMETERS	cm	0.3837 INCH
1 MILLIMETER	= 1000 MICRONS	mm	
1 MICRON	= 1/1000 MILLIMETER OR	μ	
	1000 MILLIMICRONS		NO ENGLISH
1 MILLIMICRON	= 10 ANGSTROM UNITS	$m\mu$	EQUIVALENTS
1 ANGSTROM UNIT	= 1/100,000,000 CENTIMETER	Å	

MEASURES OF CAPACITY		SYMBOL	ENGLISH EQUIVALENT
1 KILOLITER	= 1000 LITERS	kl	35.15 CUBIC FEET OR 264.16 GALLONS
1 LITER	= 10 DECILITERS	l	266.1025 CUBIC INCHES
1 DECILITER	= 100 MILLILITERS	dl	.03 FLUID OUNCES
1 MILLILITER	= VOLUME OF 1 g OF WATER AT STANDARD TEMPERATURE AND PRESSURE.	ml	

MEASURES OF MASS		SYMBOL	ENGLISH EQUIVALENT
1 KILOGRAM	= 1000 GRAMS	kg	2.2046 POUNDS
1 GRAM	= 100 CENTIGRAMS	g	15.432 GRAINS
1 CENTIGRAM	= 10 MILLIGRAMS	cg	0.1543 GRAINS
1 MILLIGRAM	= 1/1000 GRAM	mg	ABOUT .01 GRAIN

MEASURES OF VOLUME		SYMBOL	
1 CUBIC METER	= 1000 CUBIC DECIMETERS	m^3	
1 CUBIC DECIMETER	= 1000 CUBIC CENTIMETERS	dm^3	
1 CUBIC CENTIMETER	= 1000 CUBIC MILLIMETERS	cm^3	
1 CUBIC MILLIMETER	= 1 MILLILITER	mm^3	

Like the standards for mass, volume, and distance, the temperature scale which scientists use is different from the traditional Fahrenheit system. Table 5–2 shows three temperature scales in common use today. In our daily affairs, most of us use the Fahrenheit scale. Scientists use the centigrade system most frequently. It has the advantage of being based on units of ten—there are 100 degrees between the boiling point and freezing point of water. The absolute or Kelvin (K) scale is also frequently used in scientific work, for temperatures falling below the normal range encountered in most situations.

TABLE 5–2 COMPARISON OF TEMPERATURE SYSTEMS

	°F	°C	°K
BOILING POINT OF WATER	212	100	373
FREEZING POINT OF WATER	↑ 180° ↓ 32	↑ 100° ↓ 0	↑ 100° ↓ 273
LOWEST TEMPERATURE OF UNIVERSE	−459	−273	0

5-3 NORMAL CURVES AND THE ANALYSIS OF DISTRIBUTIONS

Many types of measurements in biology involve sampling small amounts of data from the vast amounts which are potentially available. For instance, it would be impossible from a practical point of view to measure the height of all the human beings in a large city in order to determine the average height of the city's population. Not only would such a procedure be time-consuming and laborious, it would also be unnecessary, since by choosing sample individuals from among the population at large, acceptably accurate results are obtained with a minimum of effort. This means that if the sampling is **unbiased** (*i.e.*, if all the individuals are not taken from one neighborhood, or from one age group), the average height calculated from a fraction of the total population should be nearly the same as that from the entire population. Thus measurements of height on 1000 adult males between the ages of 25 and 35 out of a population of 500,000 should be equivalent to that found for all adult males between 25 and 35 in the population. One of the chief problems in any sampling technique, however, is the possibility that the chosen few measurements will be **biased**— that they will not be representative of the population as a whole. There are several ways in which the degree of bias, affecting the validity of sampling data, can be determined. Some of the ways in which such data can be treated will be discussed in this section. First, however, let us consider how sampling is done, and how the raw data collected from field or laboratory measurements are converted into a useful form.

Several years ago a group of biologists set out to measure a specific characteristic, tail length, in two closely related species of deer mice (genus *Peromyscus*). They suspected from previous reports that the two populations were different in this characteristic, and wanted to see whether this difference was statistically significant. The first collecting trip in the field yielded only 15 specimens from population A. Although this was a small sample, the organisms were brought back to the laboratory and their tail lengths measured. The values that were recorded are shown in Table 5–3.

Note that for the sake of accuracy, three different observers measured each organism. The slight discrepancies in the values recorded for each organism reflect a type of error that inevitably results when measurements are made. All measurements involve some element of estimation (for example, where does one start measuring a mouse's tail?), and thus some differences will always arise. Despite the fact that measurements are objective and quantitative, they still involve human judgment, and thus the element of subjectivity. Where big discrepancies in data arise, it is important to know whether those discrepancies are a valid measure of some difference found in nature or are the result of an error in measurement. For example, observer 2 gave the value of 50.9 mm for organism 10, whereas observers 1 and 3 gave it as 60.8 and 60.7, respectively. In this case, with three observers checking each item, we would be safe in concluding that the value of 50.9 was in error. However, if we had only the data from observer 2 to work with, it would not be so easy to discard the suspect measurement. The low value of 50.9 (compared to the other 14 organisms) could be due to a mutilation (in which case it could be discarded as atypical) or a mutation (in which case it would be of statistical importance to the biologist). Now, considering for further statistical purposes only the data collected by observer 1, let us see what information these measurements can yield.

TABLE 5–3

ORGANISM NO.	TAIL LENGTH, mm		
	OBSERVER 1	OBSERVER 2	OBSERVER 3
1	60.5	60.2	60.3
2	61.0	59.9	61.1
3	62.2	62.0	63.0
4	68.1	68.0	67.9
5	60.7	60.6	60.2
6	58.3	58.4	58.5
7	66.6	66.0	66.3
8	56.7	56.6	56.6
9	62.5	62.6	62.5
10	60.8	50.9	60.7
11	58.0	58.2	58.1
12	54.5	54.5	54.5
13	56.7	56.2	56.1
14	58.9	58.8	58.7
15	60.2	60.3	60.2

There are a number of ways in which data can be analyzed and its validity checked. One important step in the analysis is to calculate the **mean** or average value of tail length for the sample at hand. The mean (symbolized $\overline{X}$) is determined by adding up (summing, symbolized by the Greek letter Σ) all the individual values (1 through 15) and dividing by the total number of values (15) in the sample. This can be expressed mathematically as

$$\overline{X} = \frac{\Sigma \overline{X}}{N},$$

where ΣX represents the sum of all the individual measurements, and N the total number in the population. Using the values from observer 1, we find that

$$\overline{X} = 60.3.$$

The mean value is useful in comparing one sample with another.

The data collected in Table 5–3 represent a survey of 15 organisms from a large natural population. The average of these 15 individuals, 60.3, may not be typical of the whole population. The smaller the sample, the greater the chance of **sampling error**—of results not typical of the whole. After all, finding the mean of a sample of data is nothing more than making a generalization, and as was shown in Chapter 3, it is dangerous to generalize from a small sample of data. It is therefore important to have as large a sample of measurements as possible. For instance, if organisms 14 and 15, as recorded in Table 5–3, had values of 40.0 (and, let us say, represented mutations), the mean would be lowered from 60.3 to 57.6. This is a large difference if one is talking about the average for the entire population. The question would then become: Are 2 out of every 15 mice mutants for tail length, or did 2 mutants just happen to turn up in the present sample?

Sampling error is a significant problem in the collection of data. A large sample helps to reduce the misleading effects of such error. The data shown in Table 5–4 show further measurements made from the original field population. How well does the mean value of 60.3 hold when a larger sample of measurements is included? (Check, using the values for observer 1.)

5-4 FROM TABLE TO GRAPH

The data in Tables 5–3 and 5–4, in their present form, yield only a small amount of information. Calculation of the mean is one step toward analysis of the results in a meaningful form. A second step is organization of the data into a graphical form that will tell us not only what the average value is, but how each measurement relates to that average.

One of the most common ways to graph data such as those given above is to construct a **histogram**. A histogram is a form of bar graph. The height of each bar measures the number of individuals. The placement of the bar along a horizontal line indicates the specific range of value for each measurement. For example, the extremes of tail length given in Tables 5–3 and 5–4 run from

52.4 to 68.0, excluding organism 85.* On this basis, nine categories, each representing a range of 2 mm, can be set up along the horizontal axis of the bar graph. Along the vertical axis the number of organisms can be measured off, from 1 to 25. The histogram is constructed by making the bar for each category correspond in height to the number of organisms in that category. The completed histogram is shown in Fig. 5–1.

TABLE 5–4

ORGANISM NO.	TAIL LENGTH, mm			ORGANISM NO.	TAIL LENGTH, mm		
	OBSERVER 1	OBSERVER 2	OBSERVER 3		OBSERVER 1	OBSERVER 2	OBSERVER 3
16	60.3	60.3	60.5	51	57.2	57.5	57.3
17	64.5	64.6	64.5	52	59.1	59.2	59.0
18	61.1	61.5	61.7	53	62.5	62.4	62.0
19	62.1	62.2	62.0	54	63.6	63.6	63.5
20	62.7	62.7	62.7	55	57.0	57.1	57.2
21	61.0	59.8	61.2	56	55.1	55.5	55.0
22	65.9	65.8	65.7	57	56.5	56.6	56.5
23	64.4	62.9	63.0	58	62.5	62.6	62.4
24	61.0	61.2	61.0	59	56.7	56.5	56.3
25	57.7	57.6	57.5	60	58.5	58.6	58.5
26	60.5	60.4	60.5	61	60.1	60.0	60.2
27	58.5	58.8	58.6	62	59.8	59.9	60.0
28	60.1	60.3	60.5	63	65.5	65.4	65.0
29	61.8	61.9	62.0	64	58.1	58.2	58.3
30	59.5	59.7	59.6	65	62.2	62.2	62.2
31	61.5	62.0	61.7	66	56.5	56.6	56.6
32	64.5	65.0	64.8	67	63.6	63.1	63.5
33	63.0	63.1	63.2	68	61.0	61.0	61.0
34	61.9	61.8	61.8	69	58.7	58.6	58.8
35	60.2	60.2	60.2	70	62.0	62.2	62.4
36	56.5	56.6	56.4	71	67.1	67.2	67.0
37	64.0	64.1	64.0	72	59.8	59.7	59.8
38	60.8	60.6	60.6	73	62.5	62.6	62.6
39	61.0	61.1	61.1	74	58.0	58.7	58.6
40	65.1	65.1	65.3	75	62.5	62.3	62.6
41	62.5	62.5	52.6†	76	65.1	65.0	65.2
42	63.0	63.0	63.0	77	58.5	58.7	59.0
43	60.1	60.4	59.9	78	62.5	62.6	62.7
44	52.5	52.3	52.4	79	56.5	56.5	56.5
45	65.5	65.1	65.7	80	63.4	63.6	63.8
46	62.0	62.1	62.1	81	54.0	54.2	54.2
47	60.8	60.7	60.7	82	62.2	62.4	62.3
48	58.8	58.7	58.6	83	64.5	64.4	64.4
49	59.1	60.0	59.5	84	57.8	58.0	57.8
50	63.1	63.0	63.2	85	49.1	49.2	49.0

---

* For the sake of this example, let us assume that organism 85 represented a mutilation rather than a mutation.
† Possible error in measurement or in recording of data.

The histogram has the advantage of showing immediately on inspection the category containing the largest number of organisms. This is called the **mode.** The histogram also has the advantage of showing immediately the way in which other categories are grouped around the mode. In Fig. 5–1, the other categories are grouped roughly evenly to both sides of the mode, and thus represent what is called a **normal distribution.**

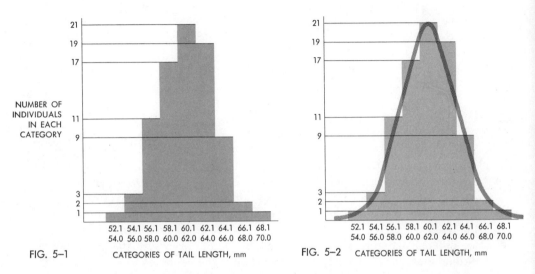

FIG. 5–1   CATEGORIES OF TAIL LENGTH, mm

FIG. 5–2   CATEGORIES OF TAIL LENGTH, mm

FIG. 5–1. Histogram of data compiled in Tables 5–3 and 5–4. Histograms show clearly the number of individuals in each category. The largest single category (in this case, 60.1–62.0 mm) is called the mode.

FIG. 5–2. Line graph of data presented in Tables 5–3 and 5–4, superimposed on the histogram of these data.

In a perfectly normal distribution, the mean and the mode are identical; however, the mode is not necessarily the same as the mean, and the two should not be confused. In principle they represent different quantities. It is true that in many of the sampling procedures which biologists deal with, the two are identical or very close together. To determine the mean accurately, however, it is still necessary to make the calculations that we discussed in the previous section.

If a line is drawn through the bars on the histogram, the data can be shown as a **line graph** (see Fig. 5–2). This graph shows a **normal distribution curve.** It is more or less symmetrical, with the value having the greatest frequency in the center and with values decreasing equally on both sides.

We have already discussed two statistics relating to graphs: the mean and the mode. On a graph, the mean is a measure of the degree to which the values fall in the middle of the range of observed values. The mode, as in the histo-gram, represents the category containing the largest number of individual items of data.

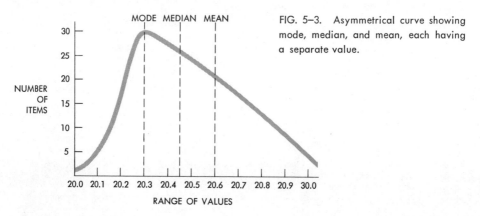

FIG. 5–3. Asymmetrical curve showing mode, median, and mean, each having a separate value.

One other statistic is of importance—the **median.** This is the value of the observation which has an equal number of observations to either side. The median always defines the recorded value that is the midpoint of a number of observations. If a distribution curve is symmetrical (*i.e.*, if it is a normal curve) the median and the mean are the same. If the curve is asymmetrical, the two are different. Figure 5–3 shows an assymetrical curve with the mean and the median marked. The median occurs between the values of 20.4 and 20.5. The mean, however, is found at 20.59, that is, nearly at the value 20.6. The mean, median, and mode thus represent different ways of describing the degree of symmetry of a curve.

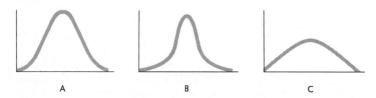

FIG. 5–4. Three curves of frequency distribution with the same mean, but different dispersion of data around the mean. The variance for curve C would be the greatest, whereas that for B would be the least.

Despite the methods developed thus far to describe the distribution of data, we still have no adequate way to estimate the *dispersion* of the observations. For example, all three curves shown in Fig. 5–4 are symmetrical and have identical means, modes, and medians. Yet it is obvious that in each case the data are dispersed quite differently around the mean.

One statistic that provides an estimate of dispersion is called the **variance,** and is symbolized by s^2. The greater the value for the variance, the more widely dispersed are the data around the mean. Of the three curves shown in Fig. 5–4, curve C obviously has the greatest variance. To relate this to our earlier example, we should expect that the variance for a distribution of tail length should be much greater in a natural population of mice than in an inbred laboratory strain. Both types of mice will show approximately normal distribution curves but the laboratory mice, because they are genetically more similar to each other, will show less variance in the dispersion of the measurements. The dispersion could also be measured by the **range,** *i.e.*, the difference between the minimum and maximum measurements in the data. But then we would know only the lower and upper limits of measurement, irrespective of how great the frequency of the various measurements might be. Variance has the advantage of taking frequency of distribution into account; for this reason, it is capable of yielding much more information about the actual dispersion of measurements around the mean.

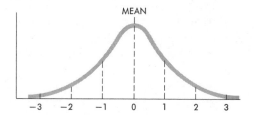

FIG. 5–5. A normal distribution curve showing the units of standard deviation marked off from the mean. If the sampling of data is valid, the standard deviation for most items of data will, in general, be within one unit on either side of the mean.

A second statistic which is of considerable value in describing dispersion of data on distribution graphs is called **standard deviation.** Standard deviation is merely the square root of the variance. It provides a way of showing the limits within which any given items in a distribution may be expected to deviate from the mean. All, or nearly all, of the observations in a sample usually lie within three standard deviations of the mean. As is shown in Fig. 5–5, 68% of the observations fall between the -1 and $+1$ values on the graph. In other words, 68% of the observations can be expected to lie within one standard deviation unit from the mean. Similarly, 95% of the observations are expected to fall within two standard deviation units of the mean, and over 99% of the observations are expected to fall within three standard deviation units from the mean.

Mean, median, mode, variance, and standard deviation all provide ways of describing characteristics of distribution curves. Since the distribution curve is one of the most common types of graph encountered in biological research, familiarity with these statistical concepts is important to the biologist. Not only do they help him to express differences between one set of data and another, they also help him to determine the validity of sample information. If the standard deviation for an item is more than three units, it is apparent that the data represent a very far spread from the mean. In addition, once the standard deviation and variance have been calculated for any sampling, the relation of each item to the mean for the whole sample can be calculated very easily. A questionable item of data can thus be checked for its uniqueness. Through use of such statistical tools, measurements can be treated to a variety of tests that help the researcher draw more information from them than is found in the raw data themselves.

5-5 CORRELATIONS

We have seen that graphs summarize a large amount of data very succinctly. Graphs are also useful in that they can show correlations between factors. For example, suppose that we release a large number of microorganisms, say *Paramecia* (singular, *Paramecium*), in a large culture dish. Assume that the central part of the dish is marked off with a circle, and all the *Paramecia* start out within this circle. If we observe the circle once an hour for 6 hours, we find that the number of individuals within the circle gradually decreases. Data for these observations are collected in Table 5–5, and graphed in Fig. 5–6.

In Fig. 5–6, it can be seen that there is definite relationship between time after release and number of *Paramecia* in the central circle. This is not unexpected. Since the organisms swim about randomly, they gradually disperse themselves throughout the entire dish; the number in the central circle thus gradually decreases.

TABLE 5–5*

NUMBER OF HOURS AFTER RELEASE	1	2	3	4	5	6
NUMBER OF *PARAMECIA* IN CENTRAL CIRCLE	104	31	16	8	5	5

* Taken from Andrewartha, H. G., *An Introduction to the Study of Animal Populations* (Chicago: University of Chicago Press, 1963), p. 214.

The graph provides more than just this information, however, for the line does not descend in a straight path, but rather in a curve of decreasing slope. The steepness of the slope between the first hour and the second indicates that more organisms left the central circle in that interval than in any other interval observed. In other words, from the graph we learn something about the *rate* at which the number of individuals in the starting circle decreases. If treated mathematically, the slope of the line gives a quantitative measure of this rate.

Such analysis of several parts of the curve gives even further information—namely, the *change* in rate. Thus we see that from a simple graph a large quantity of information can be derived.

5-6 INTERPOLATION AND EXTRAPOLATION

To **interpolate** means, literally, to "polish between." In terms of data analysis, the term refers to filling in between two items of data. In Fig. 5–6, we did not measure the number of *Paramecia* in the circle between hours 1 and 2. If we did, however, we would expect the number to be about 50, *i.e.*, on a direct line between 104 and 31. It is on the basis of this assumption that we drew the graph line connecting these two items of data. Drawing such a graph line is an example of interpolation, for it involves the estimate of an unmeasured quantity, based on the trend shown by items of data on either side of it.

Note carefully the basic assumption involved in interpolation—that the trend shown by the two items of data is representative of points between them not directly measured. There is no guarantee that this is true for any individual case; however, experience has shown that when a reasonably large sample of data is available, interpolation is a quite valid process. Since there is theoretically an infinite number of possible measurements which could be made in constructing the graph shown in Fig. 5–6, some form of interpolation is essential.

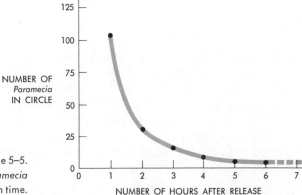

FIG. 5–6. Graph of data presented in Table 5–5. This graph shows that the number of *Paramecia* in the region where released decreases with time.

On other occasions, however, it is desirable to go *beyond* the data which are available. For example, in Fig. 5–6, we might wish to estimate the number of *Paramecia* in the circle after 7 or 8 hours. On the basis of the curve at the last two measured points (5 and 6 hours), it would be logical to extend the curve to the right in a horizontal line. At 7 hours the number of *Paramecia* within the circle would thus be 5. The process of extending the graph line beyond the measured data is an example of **extrapolation**.

In extrapolation, as in interpolation, projection is made by inference from a known into an unknown situation. In general, however, *extrapolation is less certain than interpolation*. A simple example will illustrate this point. Suppose that an oil prospector finds oil in three wells, all in a straight line with each other. By interpolation he would be reasonably certain of finding oil at a fourth well, dug somewhere on this line between the other three. He could infer that all three wells were tapping the same vein as an oil source. By extrapolation, however, the prospector would be much less certain of finding oil in a fourth well dug on an extension of the line. The vein could well end just beyond the third point, thus providing no source of oil for the fourth drilling.

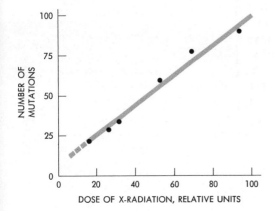

FIG. 5–7. Graph showing relative number of mutations observed in irradiated organisms against the dose of radiation received. [Adapted from B. M. Duggar, *Biological Effects of Radiation*, Chapter VI: "Statistical Treatment of Biological Problems in Irradiation," by Lowell Reed (New York: McGraw-Hill, 1936.)]

The validity of certain extrapolations is often the subject of heated scientific controversy. One of the most widely discussed topics in modern biology is based on the question of the effects which atomic radiation may have on living organisms. There is little doubt that the mutation rate in organisms is directly proportional to the amount of radiation received—as long as the dosage is above a certain level. Below that level, however, a sufficient quantity of valid data is difficult to obtain. On the basis of data available at the present time, if a graph is plotted comparing the dosage of radiation received against the number of mutations observed, a proportional relation is observable (see Fig. 5–7).

For small doses, such as those that could conceivably result from nuclear fallout, experimental evidence is still scanty, and no valid conclusions have yet been drawn. In discussing the question of banning nuclear tests, physicists have taken the viewpoint (in general) that downward from about 20 units (on the graph of Fig. 5–7) the curve drops off and runs along the base line—in other words, they contend that below this point virtually no mutations occur. Biologists, on the other hand, claim that the evidence gained so far indicates that the line should continue its downward trend on the same proportional basis—that every dose of radiation, no matter how small, would produce its proportional number of mutations. As the reader can quickly see, the question is one of extrapolation. Until more evidence on small-dosage radiation is available, no satisfactory answer can be given.

Physicists claim that their viewpoint is substantiated by many such graphs of certain physical phenomena, where careful measurement has shown that the graph line does level off as it approaches the zero point. Biologists feel that their viewpoint has validity on the basis of the measurements made at higher dosage rates. The controversy is by no means trivial, since the future development of civilization may hinge, in part, on which case proves to be more nearly correct. We bring the human survival question to bear on the problem of extrapolation to underscore the importance which lies in the *uncertainty* of this method.

Both interpolation and extrapolation show only what *most probably*, or *might* be the case; nevertheless, they are important to scientists because they are ways of thinking which human beings naturally tend to use. Both involve inferring from known information, or data, by an inductive path toward general conclusions. The conclusion in these cases does not inevitably follow—at best it is only highly probable. Yet, interpolation and extrapolation afford the scientist fairly reliable methods of prediction. Without venturing to extend his conclusions beyond the small items of data which he has collected, the scientist would be hard put to draw *any* conclusions. This would render all data meaningful only in context of the specific situation in which they were collected. All generalizations would be meaningless if we could not have some faith in our extrapolations and interpolations.

5-7 GENERALIZING POINTS ON A GRAPH

Interpolation and extrapolation represent one type of generalizing activity in which researchers engage when analyzing data. In constructing graphs, another type of generalizing also occurs. Consider Fig. 5–8. A biologist plotted the data represented by the dots in an attempt to learn whether any correlation existed between the diameter of nerve fibers and the maximum velocity at which the fiber would conduct an impulse. The data present a scatter across the graph. This is characteristic of most plots of raw data. Scattering is due to experimental error, to error in measurement, and to slight physiological differences among nerve fibers.

In constructing a graph line from these data, the investigator has two choices: (1) He can draw a line connecting every point, thus producing a zigzag effect, or (2) he can draw a single straight line which passes roughly through the midline of the scatter. The second choice is preferable in this case. Drawing a single straight line has the advantage of showing more definitely the trend which the data indicate. It is, then, a means of generalizing the data. From the graph line shown in Fig. 5–8, we conclude that there is a regular, proportional relation between the diameter of a nerve fiber and the rate at which the fiber conducts an impulse.

Not every case of scattering is best treated by generalizing the trend. In graphs of the stock market, for example, the small fluctuations from day to day or week to week may be as important for some purposes as the general trend shown over the entire year. Here is a case of having to decide when small variations in data are important and when they are not. The nature of the data and the purposes of the investigator determine to what extent graph lines may be validly used to generalize a scattered number of points. It is important to recognize, however, that in many instances such generalizations are not only permissible, but desirable.

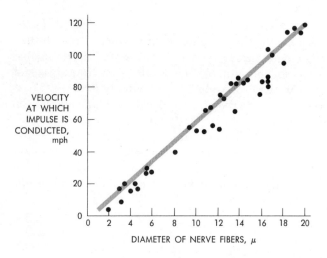

FIG. 5-8. Graph showing the relation between the diameter of nerve fibers and the rate at which they conduct a nerve impulse. The dots represent raw data and show scattering; the data are generalized with a graph line which shows the trend. Such generalizations are permissible because the data are scattered along a definite plane. (Data from *Ohio J. Science* 41, p. 145.)

5-8 THE SIGMOID CURVE

The **sigmoid curve** is found commonly in measurements of total growth in individual organisms or populations. Figure 5-9 provides an example of this type of graph line. Sigmoid curves are typical of two general areas of biological investigation: (1) studies of the growth pattern that organisms display in response to certain growth-promoting or -inhibiting substances, and (2) ecological studies where the growth of a whole population is measured.

The graph of Fig. 5-9 shows that an organism or population does not grow at an even rate. The greatest rate of growth seems to occur between the eighth and thirteenth unit of time (in this case, days). After this, the rate slows down until, around the twentieth day, the organism or population has ceased to grow at all. It is then said to remain in equilibrium.

Note that the graph can be roughly divided into four phases, each representing a different rate of growth. First comes the **positive acceleration phase,** where the organism or population is just beginning to increase; it is "getting on its feet," so to speak. Soon, however, the growth rate increases rapidly, and the curve shoots sharply upward. This is the **logarithmic phase,** so called because the increase is occurring in an exponential manner. For a single growing organism, the cells are multiplying most rapidly in this phase; for a population, the number of individuals is increasing most rapidly. Then, however, for various reasons which depend upon the situation under consideration, the growth rate slows down and enters the **negative acceleration phase.** The growth rate finally levels off in the **equilibrium phase.** Here, the number of cells produced in a single organism just equals the number which die off or, in a population of organisms, birth rate equals death rate. In this phase, the size of the individual or the population remains about the same.

Since it is a typical representation of the growth of living organisms, the sigmoid curve is one of the most common graph patterns in biology. Its occurrence is not limited to growth studies, for there are other biological situations where a sigmoid curve may be found. One example will be considered in the next section.

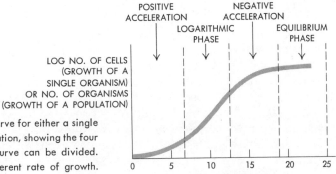

FIG. 5–9. Typical growth curve for either a single organism or an entire population, showing the four phases into which such a curve can be divided. Each phase indicates a different rate of growth.

5-9 SCALAR TRANSFORMATIONS

We have talked earlier in this chapter about the problem of choosing the proper scale for making measurements and for plotting data in a table or graph. Let us now consider a more subtle aspect of this problem. Suppose that a biologist is investigating the relation between body mass and metabolic rate* in a series of mammals. Body mass is measured in grams, and metabolic rate by the amount of oxygen consumed per gram of body mass per hour. Data on seven animals are shown in Table 5–6.

---

* Metabolic rate is the rate at which an organism releases energy (i.e., oxidizes its food supply). The most accurate measure for metabolic rate is to observe how great a volume of oxygen the organism consumes in a given period of time. The more oxygen used, the higher the metabolic rate, and vice versa.

This table shows that there is some kind of *inverse relation* between body mass and metabolic rate. An inverse relation is one in which as one quantity is increased (for example, body mass), the other decreases (metabolic rate). Thus the smallest animal has the highest metabolic rate, and *vice versa*. Unfortunately, the table itself does not show the exact nature of the relation between the two factors. If the data in Table 5–6 are plotted on a graph, however, the relation becomes much more apparent (see Fig. 5–10).

TABLE 5–6 THE METABOLIC RATES OF MAMMALS OF VARIOUS BODY SIZES

	BODY MASS, g	OXYGEN CONSUMPTION, mm^3/g-hour
MOUSE	25	1,580
RAT	225	872
RABBIT	2,200	466
DOG	11,700	318
MAN	70,000	202
HORSE	700,000	106
ELEPHANT	3,800,000	67

As the graph indicates, body mass and metabolic rate do not vary in the same proportion. Maximum oxygen consumption (mouse) is 23 times as great as minimum oxygen consumption (elephant). On the other hand, maximum mass (elephant) is over 15,000 times as great as minimum mass (mouse). The graph shows that as body mass increases from 25 to 11,700 grams (from mouse to dog), metabolic rate drops drastically. Thereafter the curve begins to level off, and eventually approaches the horizontal.

There are two difficulties with the graph as shown in Fig. 5–10. Both result from the fact that the range of values being measured is very great (for example, body mass varies from 25 grams in the mouse to 3,800,000 in the elephant). The first difficulty is that it is impossible to get all the items of data plotted on a single graph. In order to include data for the elephant, the horizontal axis would have to be three times as long as it is in Fig. 5–10. The second difficulty arises from the fact that the data for low body masses are greatly cramped. With each subdivision of the horizontal axis representing 100,000 grams, it is virtually impossible to show the difference between 25 and 225 grams. Both these difficulties are related to the type of scale used for the graph; thus the solution to the difficulties is provided by **scalar transformation**. Scalar transformation is the process by which values for data plotted on one type of scale are converted into values on another scale.

In Fig. 5–10, the data have been plotted on an **arithmetic scale**. That is, the jump between any two adjacent values (markings on the scale) contains the same number of units as the jump between any other two adjacent values.

There are as many grams represented between the 100,000 and 200,000 marks on the horizontal axis as there are between the 200,000 and 300,000 marks. By performing scalar transformation, we can convert arithmetic scales to **logarithmic** (or **exponential**) **scales.** Logarithmic scales are very useful in scientific work. In constructing such scales, the logarithm of a number is used rather than the number itself. A **logarithm** is simply an **exponent.**

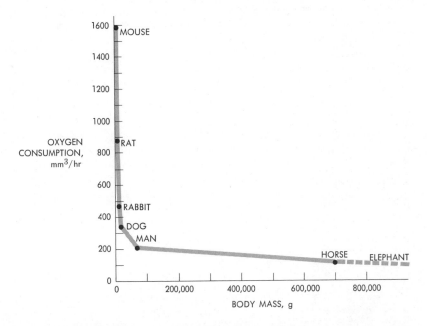

FIG. 5–10. Plot of relation between body mass and oxygen consumption for several mammals. Data are taken from Table 5–6. Both the horizontal and vertical axes show arithmetic scales.

The logarithm of the number x is simply the exponential power to which another number (called the *base*) must be raised in order to equal x. For example, 1 can be said to be the log (abbreviation for logarithm) of the number 10 (written as log 10), since $10^1 = 10$. In this example, the base number is 10 and the exponent 1. Similarly, log 100 = 2, since $10^2 = 100$; again, 10 is the base number. Log 1000 = 3, since $10^3 = 1000$. The most commonly encountered log scales are based on powers of the number 10. However, log scales can be developed using any number as the base. Since 10 is used as the base number so much of the time, logarithms with base 10 are sometimes referred to as "common" logarithms. Unless otherwise indicated, log values are assumed to be those of common logarithms.

A log scale is shown graphically in Fig. 5–11. Note that each successive unit on the log scale increases by a factor of 10, rather than by a constant increment as on an arithmetic scale. Thus the increase in arithmetic units between log-scale 1 and 2 is not equal to that between 2 and 3. From 1 to 2 would be a jump of 10 units, from 2 to 3 would be a jump of 100 units, from 3 to 4 would be a jump of 1000 units, and so on.

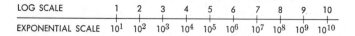

FIG. 5–11. A logarithmic scale, with 10 as the base number. On this scale, we see that a jump from 1 to 2 is much different from a jump from 2 to 3, in terms of the arithmetic numbers involved.

If we plot the data from Table 5–6 on a log scale (see Fig. 5–12), we find that we avoid the two difficulties discussed in connection with Fig. 5–10. We also find that the curved graph line becomes a straight one. (Compare Figs. 5–10 and 5–12.) Now, whenever data plotted on a graph produce a straight line, the investigator knows that some consistent relation exists between the two factors he is measuring. Thus we can see that there is a consistent relation between body mass and metabolic rate—a fact that was not so evident in Fig. 5–10. The relation has become evident only after scalar transformation. Often, data plotted on an arithmetic scale seem to show relatively less consistent relation than the same data plotted on a log or exponential scale.

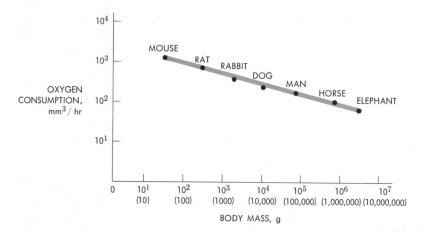

FIG. 5–12. The data plotted on an arithmetic scale in Fig. 5–10 are shown here plotted on a log or exponential scale. Exponential scales have two advantages: (1) They make it possible to plot greater extremes of data on a single graph, and (2) they show certain types of relations which are not so apparent on arithmetic scales.

The question which now arises is: Why should the dependent and consistent relation between body mass and metabolic rate show up when the data are plotted on a log scale, but not when the data are plotted on an arithmetic scale?

Living organisms have both a surface area and a volume. As the size of the organism increases, volume and surface area increase at different rates—volume increases more rapidly than surface area. This means that a large organism has a larger volume in relation to surface area than a small organism.

The significant feature of this relationship is that surface area provides the point of exchange of heat between the organism and the environment. For a large animal, the volume is large enough to compensate for the loss of heat at the surface. For a small animal, however, with proportionately much more surface area in relation to its volume, a great deal of heat is lost to the outside. The correspondingly higher metabolic rate that is characteristic of small animals releases more energy in a given period of time and some of this energy serves to keep the body temperature at a relatively constant rate.

5-10 CONCLUSION

In the present chapter we have discussed several of the most important principles in the collection and analysis of data. The design of an experiment is of crucial importance in modern science. Collection and analysis of the data resulting from that experiment, however, is the heart of scientific discovery. The conclusions, the relationships, the trends which data can be made to yield depend upon how the data are handled. It is important not only for scientists, but also for those approaching science for the first time, to become familiar with procedures used in analyzing data. We have seen in this chapter how a formidable array of numbers such as that shown in Tables 5–3 and 5–4 can yield important information. With patience and imagination, the researcher can extract much meaningful information from raw data.

EXERCISES

Answer each part of each question specifically.

1. Figure 5–13 shows the growth curve of a mushroom.
    a) What is the phase between 0 and about 15 hours called? What is probably happening within the mushroom in this phase?
    b) What is the phase between 15 and 25 hours called? Between 25 and 30 hours? Compare what is happening within the mushroom in these two stages.
    c) What is the phase from 30 hours onward called?
    d) Can you offer any hypothesis as to why the mushroom curve should level off after about 30 hours?

2. Figure 5–14 shows a comparison of wet mass to dry mass in mushrooms. Wet mass is the mass of the whole mushroom, including the water which makes up a great percentage of the bulk; dry mass is the mass of the solid material in the mushroom, with all the water evaporated.

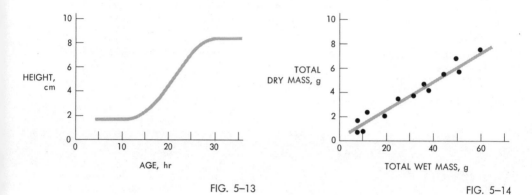

FIG. 5–13   FIG. 5–14

a) What will be the approximate dry mass of a mushroom whose wet mass is 40 g? What will be the approximate wet mass of a mushroom whose dry mass is 2.1 g?

b) Since this graph represents the relative masses *during growth*, what do we see about the amount of water which a mushroom contains as it grows larger?

c) Biologists have long been interested in exactly how a mushroom manages to grow so rapidly. One hypothesis offered was that the mushroom absorbs a great quantity of water at one stage in its growth and therefore greatly increases its size in a relatively short time. Evaluate this hypothesis in light of the graph.

3. The curves shown in Fig. 5–15 illustrate the total amount of growth in length of roots subjected to a certain experimental treatment (the units could be milliliters of poison or hormones, roentgens of radiation, etc.). The amount of this treatment increases steadily from curves A to F. From the six graphs, which of the following statements may be validly made?

a) The experimental treatment had no effect on the growth of these roots.

b) After being exposed to the experimental treatment, the roots took much longer to arrive at their maximum length.

c) Low dosages of the experimental treatment probably have little effect upon the growth of these roots.

d) The exposure to the experimental treatment resulted only in a slowing down or inhibition of root growth; there was no effect upon the final length of the mature root.

e) If we graphed a comparison of the concentration of the experimental substance used in treatment with the effect upon total growth of the roots, we would get a line of perfect positive correlation.

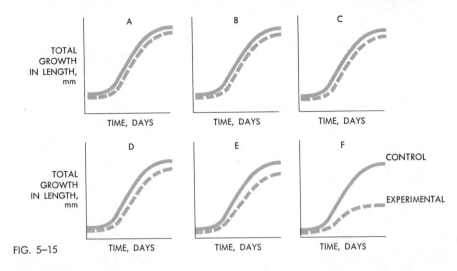

FIG. 5–15

4. The data given below are the weight in pounds and the height in inches of a group of men.

WEIGHT	147	149	153	151	155	154
HEIGHT	67	68	70	69	72	71

Which one (or ones) of the graphs in Fig 5–16 represents a reasonably correct plotting of these data? If you were writing a scientific paper on this subject, which graph would you select to illustrate your data? Why?

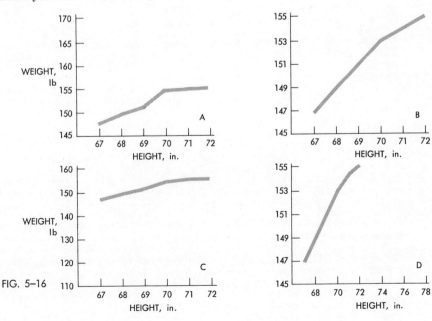

FIG. 5–16

5. Figure 5–17 is a graph* plotting growth (measured in centimeters per day) against the amount of vitamin B₆ in a nutrient solution fed to *Neurospora*, the red bread mold. *Note:* The amount of vitamin B₆ supplied in the nutrient is the *only* variable factor in this experiment. Read the lettered statements below and write on your paper: A, if the statement is in agreement with the data; B, if the statement is contradicted by the data; C, if the statement cannot be judged either valid or invalid on the basis of the evidence.

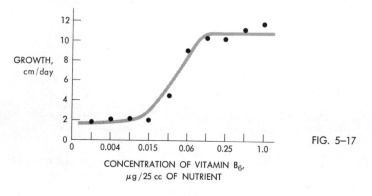

FIG. 5–17

a) There is a positive correlation between the amount of growth of *Neurospora* and the concentrations of vitamin B₆ fed to the culture.

b) A *Neurospora* culture will grow most rapidly on a vitamin B₆ concentration of from 0.03 to 0.08 micrograms per cc of nutrient fluid.

c) A *Neurospora* culture will grow more rapidly at a vitamin B₆ concentration of 1.0 than at either of the figures given above in (2).

d) Increasing the concentration of vitamin B₆ to 1.5 micrograms per 25 cc of nutrient solution would probably not increase the rate of growth to any appreciable degree.

e) Vitamin B₆ is a building block necessary for the construction of some necessary compound in the living *Neurospora*.

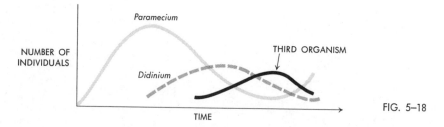

FIG. 5–18

6. Figure 5–18 is a graph showing population cycles in a laboratory culture dish of three organisms. The original culture dish contained the protozoan *Paramecium;* a second

---

* Graph taken from G. W. Beadle and E. L. Tatum, "The Genetic Control of Biochemical Reactions in *Neurospora*." *Proceedings of National Academy of Sciences* **27** (1941), 499–506. Used by permission.

organism, *Didinium*, slightly smaller than *Paramecium*, was introduced into the culture dish when *Paramecium* was multiplying rapidly. A third organism, which happens to feed exclusively on *Didinium*, was introduced into the culture when *Didinium* first began to decline.

a) Offer a hypothesis to explain the data shown in the graph. Include a discussion of why the *Paramecium* curve rises a second time, and some ideas as to the cause-effect relationships shown here. Be as specific and complete as possible.

b) What would be the probable effect on the population of *Paramecium* and the third organism if *Didinium* were removed?

c) What would be the probable effect on the population of *Paramecium* and *Didinium* if the third organism were removed?

SUGGESTED READINGS

LINN, CHARLES, *Probability and Statistics* (Columbus, Ohio: American Education Publications, 1964). A highly readable discussion of this subject, with sections of considerable importance for the treatment of biological data.

MORONEY, M. J., *Facts from Figures* (New York: Penguin Books, 1958). This book is very well written. It presents some interesting problems in the handling of data, and is easy to read.

MOSTELLER, F., R. E. K. ROURKE, AND G. B. THOMAS, *Probability and Statistics* (Reading, Mass: Addison-Wesley, 1961). This is a standard textbook on theories of probability and statistical analysis.

TIPPETT, L. H. C., *Statistics* (New York: Oxford University Press, 1944). This is one part of the "Home University Library of Modern Knowledge." It is very elementary, providing a good introduction for those who have no previous knowledge about statistics or the handling of data.

WALLIS, W. A., AND H. V. ROBERTS, *Statistics: A New Approach* (Glencoe, Ill.: The Free Press, 1956). Although written by professional statisticians, this book represents a very readable and interesting account of various methods and problems involved in statistical analysis.

YOUDEN, W. J., *Experimentation and Measurement* (Washington: National Science Teachers' Association, 1962). This is a simple, well-written introduction to the problems of setting up experiments, collecting data, and analyzing them. Some mathematics is introduced, but is carefully explained.

PART II

This part is divided into three sections. Section 1 consists of an analysis of a scientific research paper, somewhat similar to the analysis of Spallanzani's paper on pages 29–33. Section 2 gives a sample controversy in science. Section 3 contains a similar controversy, but this time in an area which demonstrates the complex interrelationships which may arise between science and fields generally part of the social sciences and humanities.

A 100 million year old ant. Courtesy of Harvard University News.

ANALYSIS OF A SCIENTIFIC
RESEARCH PAPER

THE FIRST MESOZOIC ANTS*

Edward O. Wilson, Frank M. Carpenter, and William L. Brown, Jr.†

Abstract. Two worker ants preserved in amber of Upper Cretaceous age have been found in New Jersey. They are the first undisputed remains of social insects of Mesozoic age, extending the existence of social life in insects back to approximately 100 million years. They are also the earliest known fossils that can be assigned with certainty to aculeate Hymenoptera. The species, Sphecomyrma freyi, is considered to represent a new subfamily (Sphecomyrminae), more primitive than any previously known ant group. It forms a near-perfect link between certain nonsocial tiphiid wasps and the most primitive myrmecioid ants.

(*An abstract allows the reader of the journal to obtain a brief overview of the paper. Among other things, it enables the scientist-reader to decide whether the information contained within the paper is of direct, indirect, or no pertinence to his own field of research interest. Thus, in essence, an abstract can be a valuable time saver.*)

* *Science*, 26 June, 1967. Reproduced by permission.
† Wilson and Carpenter, Museum of Comparative Zoology, Harvard Univ., Cambridge, Mass. Brown, Department of Entomology, Cornell Univ., Ithaca, New York and Museum of Comparative Zoology, Harvard Univ., Cambridge, Mass.

Until now the earliest known fossils of ants, and of social insects generally, have been Eocene in age (1). Large assemblages of ant species, most belonging to living tribes and even genera, occur in the Baltic Amber (Oligocene), the Sicilian and Chiapan ambers (Miocene), and the Florissant and Ruby Basin shales (Miocene) (2). The diversity of these faunas and the advanced phylogenetic position of many of their elements have long prompted entomologists to look to the Cretaceous for fossils that might link the ants to some ancestral nonsocial wasp group, but until now, with one doubtful exception, no relevant fossils have turned up.

The exception is the hymenopterous forewing described by Sharov (3) as *Cretavus sibiricus*, from the Upper Cretaceous of Siberia. This wing is rather similar to that of the wasp family Plumariidae, and also approaches a reasonable possible precursor pattern for the venations of known primitive ants. However, we have no guarantee that venational characters evolved concordantly with other, more truly diagnostic body characters, so we cannot even regard it as certain that *Cretavus* is an aculeate.

Cretaceous amber from Canada and Alaska contains a moderate number of insects (4), but no ants or aculeate Hymenoptera of any kind are present among them (a fact now suggesting that the Canadian amber, which has never been precisely dated within the Cretaceous, may have been formed in an earlier part of the period). Amber securely dated to the lower part of the Upper Cretaceous is fairly common from Maryland to New Jersey in de-

Here the authors of the paper state the fact that the oldest fossils of social insects have been dated only back to the Eocene. The great diversity of these forms and their complexity indicates that more primitive forms would be expected in periods predating the Eocene (i.e., the Cretaceous). Only one wasplike forewing from the upper Cretaceous is known, however. The accepted hypothesis links the nonsocial tiphiid wasps with the primitive social ants. This hypothesis predicts the existence of an ant with primitive wasplike body features.

The prediction of the former existence of an ant with wasplike characteristics is dramatically borne out—the hypothesis is supported by the discovery of these fossils.

posits of the Magothy Formation, but until recently almost no insect inclusions had been reported. In 1965, Mr. and Mrs. Edmund Frey (5), mineral collectors of Mountainside, New Jersey, found a lump of amber in clay of the same formation at the base of seaside bluffs at Cliffwood, New Jersey. The fragile lump broke into pieces, and two of these bear insects, including two well-preserved worker ants.

The two specimens appear to belong to the same species; one is shown in the cover photograph.* We judge this species, *Sphecomyrma freyi*, to be by far the most primitive member of the Formicidae (ants) yet discovered. It is sufficiently removed from all other ants to be received into a distinct subfamily, the Sphecomyrminae. The most distinctive morphological features, and our assessment of their phylogenetic significance, can be summarized as follows.

1) The head capsule resembles that of a generalized aculeate wasp or ant. The clypeus and frontal carinae are antlike, but are of such simple conformation as not to depart significantly from these structures in some aculeate wasp groups. We regard the large, convex form of the compound eyes and their placement near the center of the sides of the head as primitive characters for aculeates generally. The presence of three large ocelli is certainly primitive.

The authors proceed to provide the reader with the necessary factual material on which to evaluate the find and its significance to the hypothesis.

2) The mandibles are short, curvilinear, and bidentate, and closely resemble those of certain species of several existing aculeate wasp families.

---

* *Editor's note:* See photograph at beginning of article.

3) The antennal funiculi are long and filiform, a trait more wasplike than ant-like. The antennal scapes (basal segments) are elongate, a characteristic of ants generally but exceptional among other aculeates; still, the scapes are shorter than is usual for worker ants.

4) The alitrunk (thorax + propodeum) is more completely sutured, and therefore more primitive, than that of any other worker ant, and is almost identical with that of the wingless females of the tiphiid genus *Methocha*. Prothorax, mesothorax, and metanotopropodeum are separated each from the next by two complete and possibly flexible sutures; and the mesonotum is composed of well-defined, convex scutum and scutellum, separated by a narrow sunken area. In fact, the only major alitruncal difference from *Methocha* is the presence in *Sphecomyrma* of apparently well-developed metapleural glands, which are peculiar to the Formicidae.

5) The single-segmented petiole, narrowly constricted behind, is an ant character state; the absence of a constriction in the gaster and the presence of a well-developed, extrusible sting are states shared by most wasps and primitive myrmecioid ants.

6) The legs show two character states that we have long regarded as primitive for ants: two spurs on each tibial apex of the middle and posterior legs, and toothed tarsal claws.

In summary, *Sphecomyrma* presents a mosaic of wasplike and antlike character states. There are nevertheless enough truly antlike traits to place *Sphecomyrma* within the Formicidae,

Note the authors' continued stress on the fossils' wasplike anatomical characteristics.

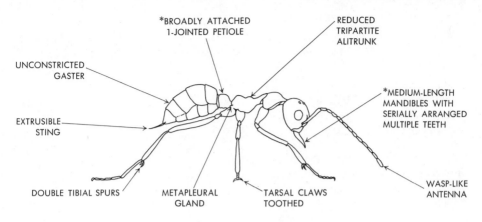

*BROADLY ATTACHED
1-JOINTED PETIOLE

REDUCED
TRIPARTITE
ALITRUNK

UNCONSTRICTED
GASTER

*MEDIUM-LENGTH
MANDIBLES WITH
SERIALLY ARRANGED
MULTIPLE TEETH

EXTRUSIBLE
STING

DOUBLE TIBIAL SPURS

METAPLEURAL
GLAND

TARSAL CLAWS
TOOTHED

WASP-LIKE
ANTENNA

PREVIOUSLY HYPOTHESIZED ANCESTOR

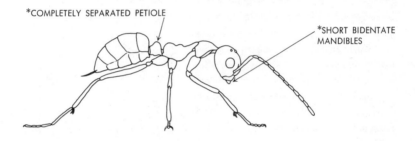

*COMPLETELY SEPARATED PETIOLE

*SHORT BIDENTATE
MANDIBLES

SPHECOMYRMA

Fig. 1. A comparison of the main features previously hypothesized by the authors to characterize the external morphology of the ancestral ant, and *Sphecomyrma* itself. The minor details of body form are arbitrarily made the same. In the drawing of *Sphecomyrma*, the starred character states indicate where our phylogenetic hypothesis proved in error.

where the most similar (but still quite different) forms are the living myrmeciine *Nothomyrmecia macrops* of Australia and the primitive aneuretine Dolichoderinae, such as *Paraneuretus* and *Protaneuretus*, of Oligocene age, described by Wheeler (2). These are primitive forms in the myrmecioid complex (6).

It is interesting to compare our earlier conception of the archetypal ant with the actuality presented by *Spheco-*

Now the original hypothesis, though dramatically supported in the main, must be modified to conform to the new

myrma. This is done in pictorial form in Fig. 1. It can be seen that our vision of what was yet to be revealed differs from *Sphecomyrma* in only one essential respect: we guessed that antlike mandibles evolved before the antlike "waist" (petiole), but the reverse actually proved to be the case.

Compared with living wasp genera, *Sphecomyrma* appears to come closest to the tiphiid genera *Methocha* (Methochinae) and *Rhagigaster* (Thynninae) (7). One interesting aspect of the morphology of *Sphecomyrma* is that in "ant characters" it does fall so close to the myrmecioid complex of genera, yet bears so little resemblance to *Amblyopone* and other genera of the Ponerinae previously regarded as nearly as primitive as the myrmecioids. The possibility is thus raised that divergence between myrmecioid and poneroid lines may already have taken place by the time *Sphecomyrma* lived. However, the presence of the complex metapleural gland in *Sphecomyrma* and all other primitive ants speaks for a monophyletic origin of the Formicidae from tiphiid ancestors. The function of the metapleural gland is still unknown, but if it turns out to mediate some phase of social behavior, then monophyletic origin of social life would be strongly implied for the ants as we know them. These new considerations are incorporated into a cladogram of the ant subfamilies (Fig. 2).

Finally, the origin of social life in the insects has now been put back from the Eocene, about 60 million years ago, to the middle or lower part of the Upper Cretaceous, about 100 million years

known data. Note the authors' use of the word, "guessed." An hypothesis is often referred to as "an educated guess."

That is, the presence of the metapleural gland supports an hypothesis proposing a monophyletic origin of the social ants from the nonsocial wasps.

The new discovery forces a modification of the evolutionary time scale postulated prior to the new discovery.

ago. It may be true that social life in insects is not much older than that. *Sphecomyrma* is evidently only a little changed from tiphiid wasps, and it is possible that this relatively slight transformation indicates a correspondingly short period of social evolution. Perhaps as more hymenopteran fossils become available from the New Jersey and similar ambers, new light will be shed on the origin of the ants.

A fuller account of *Sphecomyrma* and its phylogenetic implications, together with a formal taxonomic description, is published elsewhere (8).

The hope for more such discoveries leading to further modifications and refinements of the hypothesis is indicated by the authors.

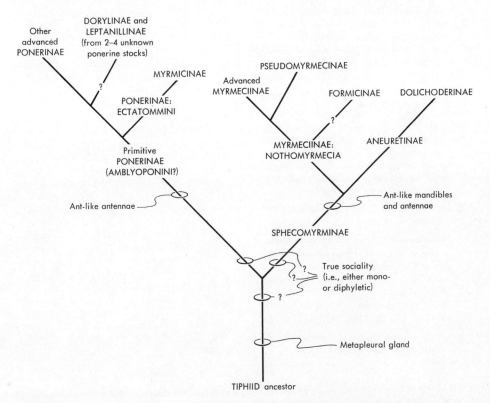

Fig. 2. A new hypothetical cladogram of the ant subfamilies taking into account the morphology of *Sphecomyrma*.

References and Notes

1. The oldest Eocene ant fossil is *Eoponera berryi*, based on a forewing from the Wilcox Clay of Tennessee; F. M. Carpenter, *J. Wash. Acad. Sci.* **19**, 300 (1929).

2. The Baltic Amber ants were monographed by W. M. Wheeler [*Schrift. Phys.-ökon. Ges. Königsberg* **55**, 1 (1914)]; and the Florissant ants by F. M. Carpenter [*Bull. Mus. Comp. Zool. Harvard* **70**, 1 (1930)].

   W. L. Brown, Jr. (unpublished) has examined the few available ant fossils from the Ruby Basin (Montana) shales and found them to match the dominant Florissant species; he has also cursorily examined the ants of the Chiapas Amber and found them related to those of the Florissant and the existing tropical Mexican faunas.

3. A. G. Sharov, *Dokl. Akad. Nauk* **112**, 943 (1957). For a comparison with wings of ants and Plumariidae, see W. L. Brown, Jr., and W. L. Nutting, *Trans. Amer. Entomol. Soc.* **75**, 113 (1950).

4. F. M. Carpenter, J. W. Folsom, E. O. Essig, A. C. Kinsey, C. T. Brues, M. W. Boesel, H. E. Ewing, *Univ. Toronto Stud. Geol. Ser.* **40**, 7 (1934).

5. We gratefully acknowledge the splendid cooperation of Mr. and Mrs. Frey, as well as the intermediary aid of Dr. Donald Baird of Princeton University and Mr. David Stager of the Newark Museum.

6. W. L. Brown, Jr., *Insectes Sociaux* **1**, 21 (1954).

7. We acknowledge the aid of H. E. Evans, who gave us the benefit of extensive comparisons of *Sphecomyrma* characters with those of various wasp genera. In classifying tiphiids, we have arbitrarily followed the system of V. S. L. Pate, *J. N.Y. Entomol. Soc.* **55**, 115 (1947).

8. E. O. Wilson, F. M. Carpenter, W. L. Brown, Jr., *Psyche*, in press.

An annotated reference list is provided so that the source of statements made in the article and the data on which the statements are based can be checked.

Figure 2 shows a diagram of evolutionary relationships existing between wasps and ants. This chart is a revision of a previous one, made before the discovery of the Sphecomyrma fossils. The question marks emphasize areas where speculation is particularly high because of a lack of hard data. The chart shown above would be only a very small portion of the phylogenetic charts shown on pages 578–579 of *The Study of Biology.*

Thus, we have here a situation somewhat similar to the discovery of a living Coelocanth off Madagascar in 1938. *Sphecomyrma freyi* is about as perfect a "missing link" as could possibly be found. The discovery lends further support to the validity of the indirect methods used by biologists to establish evolutionary relationships between extinct and living animal and plant species.

One other point might be mentioned here. This example was purposely chosen because it nicely demonstrates that the If . . . then . . . deductive format can be applied to observational as well as experimental aspects of science. Quite obviously, no experiment was performed in this case. Yet, predictions stemming from a previously proposed hypothesis were tested just the same.

CONTROVERSY IN SCIENCE

The following paper demonstrates the use of instruments commonly found in the laboratories of physical scientists in the study of a living system—an amoeba. A modern biologist must be thoroughly familiar with such instruments, their special strengths as well as their limitations. Finally, as often as not, the performing of an important experiment is frequently dependent upon the imagination of the scientist in designing the necessary tools and techniques. The use of the instruments shown in Fig. 1 is an example of such imagination. Note as you read that it is not necessary to understand all of the content of this paper in order to appreciate the work carried out by the authors.

However, this particular paper was selected because the conclusions drawn by the authors were challenged by other workers engaged in the same area of investigation. Such challenges are not at all rare in science. Their occasional occurrence is a reassurance that modern science retains its self-critical nature; that, while today's individual scientist may be as subject to error as any man, his mistakes will be corrected by others before very long.

Whether the authors of this paper or their critics were correct is immaterial here. It seems evident that further work will have to be done to ascertain where the "truth" lies. Thus, though no verdict is necessary here, the reader may enjoy playing the role of an impartial judge.

OPTICAL DIFFERENTATION OF AMOEBIC CYTOPLASM AND ENDOPLASMIC FLOW*

W. R. Baker, Jr. and J. A. Johnston, Jr.†

Abstract. Optical activity differentiating the flowing and nonflowing amoebic cytoplasm was detected. This evidence indicates molecular alignment in the flow stream and can be used to provide data on the direction of alignment. The results were obtained by utilizing a dynamic polarized-light detection system which is sensitive only to specimens which possess a preferred axis.

The application of engineering-systems techniques has led to the development of a system capable of detecting low-level optical activity in living specimens which has heretofore escaped detection (*1, 2*). Optical activity is defined here as any polarized light phenomenon. The system was developed to study the mechanisms of amoeboid motility and is now capable of differentiating cytoplasm optical activity found in the flowing endoplasm from that of the stationary ectoplasm. However, the present method of storing this information (storage cathode ray tube) limits analysis and quantification of the optical activity data. Extension of the capability of the system is required to substantiate the type of optical activity and to quantify it.

This work was directed toward the development of a system with which the material structure of an amoeba or any macromolecular substance in solution could be characterized by the orientation of its chemical bonds. The structure characterization is achieved by utilizing the changes in the polarized light absorption properties of the substance (*3*). The system makes use of a monochromatic, collimated, and polarized light source (a gas laser) and two electrooptic light modulators (EOLM's) to achieve a dynamic system with a rapidly oscillating polarized beam.

Placing the second EOLM between the specimen and a fixed analyzer enables the system to filter out all optical activities that are independent of their orientation with respect to the polarized light source. The system output information is therefore characteristic of the type of optical activity the specimen possesses and is a function of the EOLM modulation frequency and the orientation of the specimen. The system output is filtered and applied to the z-axis input of a storage tube oscilloscope. The specimen is moved under the polarized light beam by a dual servo system connected to the position control of the microscope stage. A position pick-off from the servo system is applied to the horizontal and vertical axes of the storage tube. The intensity of the storage tube beam can be adjusted so that it stores only optical activity in-

* *Science*, 14 February, 1967. Reprinted by permission.
† Biomedical Engineering Laboratory, Vanderbilt Univ., Nashville, Tenn.

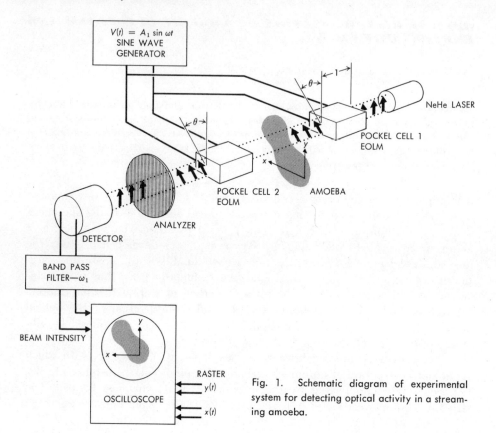

Fig. 1.   Schematic diagram of experimental system for detecting optical activity in a streaming amoeba.

formation, that is, the background noise of the system is clipped. The experimental apparatus is shown in Fig. 1.

Figure 2 is a photographic record (photograph of a portion of the storage tube) of differential optical activity of the flowing and nonflowing cytoplasm. At present this data permits comparisons of the regions of molecular alignment in the flow stream (bright regions on the photograph) and those of random alignment (dark regions on the photograph).

Further work is needed to determine the direction of molecular alignment in the flow stream to substantiate the type of activity data and compare the point of maximum intensity of the polarized beam with the angle the beam makes with the flow channel. The analysis of the optical activity data can be achieved by considering the Jones matrices (4) for the optical elements to determine the Jones matrix for the specimen optical activity. The direction of molecular alignment in the flow stream should cast some light upon the reasons for the ordering of molecules in the flow stream. From the work of Allen *et al.* (1) it is seen that the existence of a preferred axis may arise as a result of flow or stress applied to the substance.

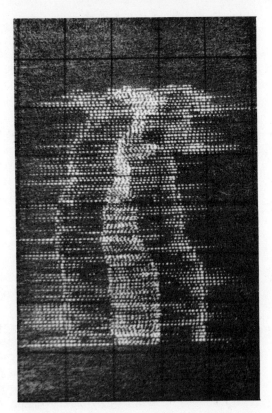

Fig. 2. Photograph of a portion of the storage tube for a scan of a *Chaos chaos* pseudopodium. Beam diameter, 20μ; light, λ, 632.8 mμ; raster size, 610μ by 800μ. Ringing of the beam modulating information at membrane and flow channel results from low damping in the tuned amplifier.

REFERENCES AND NOTES

1. R. D. ALLEN, D. W. FRANCIS, H. NAKAJIMA, "Pseudopodial birefringence and the motive force of ameboid movement," *Abstracts Biophysical Society 9th Annual Meeting* (1965), p. 151.

2. R. D. ALLEN, J. BRAULT, R. D. MOORE, *J. Cell Biol.* **18**, 112 (1963).

3. S. BEYCHOK, *Science* **154**, 1288 (1966).

4. W. A. SHURCLIFF, *Polarized Light* (Harvard Univ. Press, Cambridge, Mass., 1962).

5. Supported in part by the AEC, and the Olin Trainee Programs.

The conclusions drawn by Baker and Johnston were immediately challenged by another worker in the field.*

Baker and Johnston (*1*) offer photographic evidence that they have detected "optical activity" in living amoebas. The term "optical activity," which has traditionally meant optical rotation, is redefined by the authors to include

* *Science*, 17 May, 1967. Reprinted by permission.

"any polarized light phenomenon." However, their "dynamic polarized light detection system" is, in fact, incapable of differentiating among several phenomena detectable with polarized light (phase shifts due to birefringence, optical rotation, linear or circular dichroism, refraction absorption, surface depolarization, and light scattering). The photograph shown in Fig. 2 could have resulted from a mixture of any of these "optical activities."

In optical analysis one should identify the type of light-matter interaction observed in specimens and then proceed to quantify this interaction. Usually identification and quantification are accomplished simultaneously by null compensation for phase shifts or rotation of the plane of polarization (2). Another electronic detection technique using phase modulation referred to by the authors (3) does, in fact, use the null method through automatic compensation of rotation, phase shifts due to birefringence, and dichroism. The null compensation method, in any event, separates the phenomena detectable only in polarized light from absorption refraction and light scattering, properties that are not only detectable in polarized light but are frequently so strong that they mask birefringence and other polarized light effects.

The method of Baker and Johnston not only lacks selectivity but is apparently incapable of quantifying any single optical phenomenon. Nevertheless, the authors claim their "... system capable of detecting low-level optical activity in living specimens which has heretofore escaped detection. . . ." To be convincing, such a claim should be accompanied by data indicating the peak-to-peak or root mean square noise level for the phenomenon being registered. At present the indications are that Baker and Johnston have only transferred an ordinary light microscopic image to a storage cathode-ray tube. Any serious attempt to interpret such an image in molecular terms should be classified as "inference microscopy."

R. D. ALLEN*

*Department of Biological Sciences,
State University of New York,
Albany*

REFERENCES

1. W. R. BAKER, JR. and J. A. JOHNSTON, JR., *Science* 156, 825 (1967).

2. H. G. JERRARD, *J. Opt. Soc. Amer.* 38, 35 (1948); H. S. Bennett, in *McClung's Handbook of Microscopical Techniques*, R. M. Jones, Ed. (Hoeber, New York, ed. 2, 1939), p. 591.

3. R. D. ALLEN, J. W. BRAULT, R. M. ZEH, in *Advances in Optical and Electron Microscopy*, R. Barer and V. E. Cosslett, Eds. (Academic Press, London, 1966), vol. 1, p. 77; R. D. Allen. J. Brault, R. D. Moore, *J. Cell Biol.* 18, 112 (1963).

---

* The fact that the names Baker and Allen appear in this controversy is purely coincidental.

*Another scientist also criticizes Baker and Johnston on slightly different grounds.**

The most important point I wish to make regarding Baker and Johnston's report (*1*) is that their Fig. 2, which is presented as showing the "differential optical activity of the flowing and nonflowing cytoplasm" in a pseudopod of *Chaos chaos*, is, in fact, merely a doubled image of the pseudopod. The two images have been laterally displaced by about half the maximum width of the pseudopod. The area where the two images overlap appears brighter than its surroundings and is the region of the picture that Baker and Johnston identify as the flow channel. When the individual images are considered, no clear-cut demarcation of a flow channel is visible. The cause of the doubled image cannot be identified from the information in the report, but it could be caused by spurious reflections in the optical system or by an electronic echo in the storage oscilloscope.

Second, the experimental system they describe in their text and Fig. 1 seems to have both electrooptical light modulators (EOLM) driven in phase from the same source. If their figure is correct in this respect and shows all of the significant parts of the system, except the microscope optics, then the two modulators are functionally equivalent to a single unit having properties equal to the sum of the properties of the units shown. In this case, if the band pass filter is in fact a narrow band filter tuned to pass only signals at the EOLM driving frequency, the system will be able to detect birefringence when the birefringent axes of the specimen are not aligned with the plane of polarization of the incoming beam or the analyzer; but it will not respond linearly and will not determine the sign or the azimuth angle of the birefringence detected. Other polarized light phenomena may be detected under favorable circumstances but cannot be distinguished from birefringence without additional components in the system. Successful systems with electrooptical light modulators for automatic measurement of birefringence and other polarized light phenomena have been described by Takasaki (*2*) and by Allen *et al.* (*3*).

GORDON W. ELLIS

Department of Biology,
Univ. of Penn.
Philadelphia

REFERENCES

1. W. R. BAKER, JR. and J. A. JOHNSTON, JR., *Science* 156, 825 (1967).

2. HIROSHI TAKASAKI, *J. Appl. Phys.* (*Japan*) 28, 164 (1959); *ibid.* 29, 105 (1960); *ibid.* p. 468; *ibid.* 30, 40 (1961); *ibid.* p. 657; *J. Opt. Soc. Amer.* 51, 462 (1961); *ibid.* p. 463; *ibid.* p. 1035; *ibid.* p. 1146; *ibid.* 52, 718 (1962).

3. R. D. ALLEN, J. BRAULT, R. D. MOORE, *J. Cell Biol.* 18, 223 (1963); R. D. Allen, J. Brault, R. M. Zeh, in *Advances in Optical and Electron Microscopy.* R. Barer and V. E. Cosslet, Eds. (Academic Press, New York, 1966), vol. 1.

---

* *Science*, 29 May, 1967. Reprinted by permission.

As is the case in any reputable journal, the original authors are given the opportunity to respond.*

In our paper, we pointed out in several instances that as a result of our method of data collection, the storage cathode-ray tube, our system was, "in fact, incapable of differentiating among several phenomena detectable with polarized light. . . ." We also stated that our work will be directed toward substantiating the type of optical activity and quantifying it. Again we define optical activity as any polarized light phenomenon, information which Allen said could have resulted from a mixture of any of the possible types of "optical activities" detectable with polarized light.

Although our system does not now permit substantiation and quantification of the "optical activities," it is capable of detecting the presence of "a mixture of any of the possible types of optical activity" across amoebic pseudopodia, which as far as we know has heretofore escaped detection.

The conversion of our data from analog to digital form will allow digital filtering techniques to be used in conjunction with a Jones calculus analysis of our system elements. This type of analysis will permit the determination of the specimen matrix necessary to give the recorded system differentiation in a point fashion across the specimen. The specimen matrix specifies the type of optical activity present within the specimen. This method of handling our dynamic scan data will allow substantiation and quantification not possible with the storage cathode-ray tube.

In reply to Ellis' comment regarding the electrooptical light modulator (EOLM) voltage sources, it should be pointed out that the two EOLM's are driven by voltages that are 180 degrees out of phase. Figure 1 of the report was intended to convey this point.

A doubled image of the pseudopod, as described by Ellis, could occur only as a result of backlash between the servo-driven microscope stage and the position-indicating potentiometer. However, a doubled image resulting from servo backlash should result in the brighter portions of alternating scan lines being displaced from each other. This phenomenon is not evident from our scan photographs.† At the time of the experiment the backlash was examined and found to be negligible. A method of verifying the effect of backlash, if any, is to orient the pseudopod parallel to the suspected axis of the backlash. Such an experiment is scheduled.

Optical information of these reported observations is obscured by noise requiring vast data to be analyzed. Therefore, computer determinations of

------

* *Science*, 18 August, 1967. Reprinted by permission.
† *Authors' note:* That is, Ellis' contention predicts the displacement of the brighter scan lines, according to Baker and Johnston. The prediction is not verified by the scan photographics.

the types of optical phenomena by Jones matrices will make it possible to treat with ease the behavior of complex polarizer-retarder combinations of this dynamic system.

W. R. BAKER, JR.
JOE A. JOHNSTON, JR.

Biomedical Engineering Laboratory,
Vanderbilt Univ.,
Nashville, Tenn.

While, occasionally, such disputes between scientists may become quite heated (in the case of the embryologists Roux and Driesch, the arguments became quite bitter and personal), usually they remain balanced and impersonal. The possibility of their occurring is one of the greatest strengths of science. It keeps a scientist "honest"; he knows that his experimental data may be checked at any time. Further, the more "fame" his hypothesis obtains, the more apt it is to be tested.

ANOTHER CONTROVERSY

The following article appeared in the October 16, 1964, issue of Science, the official journal of the American Association for the Advancement of Science (AAAS). The article and the exchange of letters which follow it, provide a superb example of several misconceptions concerning science and scientists. Foremost among these misconceptions is the idea that scientists are always objective, and that they base their judgments only on a dispassionate analysis of experimental data and not on preconceived notions or judgments.

The article also demonstrates, perhaps, the need for persons holding public office to become familiar with the nature of scientific investigation in terms of its potentialities and its limitations.

Read the article carefully and make your own judgment as to its worth. As you read, try to spot areas in which you feel its arguments are weak or strong.

RACIAL DIFFERENCES AND THE FUTURE*

Dwight J. Ingle†

This article is a review of a few of the problems related to the struggle of minority ethnic groups, especially the American Negroes, for equal rights and advancement. It considers possible genetic and environmental bases of the problems, tenable means of achieving equal rights with a minimum of conflict, and the possible replacement of weakly effective efforts to alleviate biosocial problems by methods which would prevent their occurrence; and it makes a plea for

---

* *Science,* 16 October, 1964. Reprinted by permission.
† Dr. Ingle is professor of physiology at the University of Chicago.

freedom of debate and inquiry. All that follows is open to debate and criticism; this is an expression of ideas that is intended to be heuristic, not self-validating or fully documented. I use the word "race" in its popular sense, recognizing that all ethnic groups represent mixed origins and that there is no known physical or behavioral trait which is found exclusively in one "race."

The struggle by individuals and groups throughout the world for special rights and privileges is opposed more strongly than ever before by faith in man's worth as an individual and in his freedom to pursue self-fulfillment and happiness as indispensable goals of life. But one individual's drive for self-fulfillment may conflict with another's; hence society adopts laws, customs, and ethics which attempt to define rights and freedoms and thereby guide human conduct.

Most scientists accept the principle of equal legal and moral rights for the individual regardless of race or religion and support the right of each individual to advance according to his abilities, drives, and behavioral standards. But the equality of man is a social, legal, and ethical, rather than a biological concept, for among living things nothing is equal to or identical with anything else. Even if we were to achieve equal civil rights and equal opportunities, complex problems would remain.

Some ethnic groups, the Negroes especially, are handicapped by a substandard culture. Centuries of slavery and racial discrimination have left some, though not all, Negroes with a cultural handicap which begins to be transmitted from adult to child in the earliest formative years. Do genetic handicaps occur more frequently among Negroes than among other ethnic groups? Racists claim that the Negro race is genetically inferior to other races in intelligence, while equalitarians claim that all races are equally endowed with intelligence. Both groups support their respective dogmas by spurious argument and emotionalism. Although it is common to speak and write of intelligence as a unitary quality of mind, it is surely complex (*1*) and is indirectly and imperfectly measured by standardized tests. Both racists and equalitarians claim special knowledge about the relative importance of native endowment and of environment in determining the level of intelligence. The conventional wisdom of the times is that biological differences among races are of no significance from the standpoint of social action. The climate of opinion in our courts, universities, and public press is not favorable either to further inquiry into the question or to debate of the issues. Even those who recognize that the question is unresolved claim that science must stand aside in the struggle for social values.

The problems of other American minorities, such as several Asian groups and the Jews, each of which is subject to segregation, bias, and discriminatory laws, have largely been solved. Each group has a cultural heritage which has facilitated self-fulfillment and successful competition with other groups. Is the average genetic endowment of these groups better than that of the Negro? Despite discrimination against these groups, on the average they perform better than Negroes in the classroom and on objective tests of intelligence and achieve-

ment. Many individuals among other minority groups, such as the American
Indians and Puerto Ricans, are underprivileged, as are a substantial number of
native-born whites who do not suffer from racial prejudice.

The Biology of Race

Several points relevant to the biology of race seem clearly established. (i) There
is extensive overlap in the intelligence of whites and Negroes; the generalization
made by racists that all whites are superior to all Negroes is false. (ii) The
scholastic and intelligence-test performance of Negroes is on the average poorer
than that of whites. (iii) The abilities, drives, and behavioral standards of some
Negroes have led to high achievement in business, law, government, education,
and the arts and sciences; many Negroes are good schoolmates, good neighbors,
and good citizens. Conversely, indolence, incompetence, and poor citizenship
can be found among all racial groups in America, more frequently among
Negroes than among whites and Asians. (iv) Race and color are not valid
criteria for judging the worth of an individual. Whatever criteria are needed
to judge the individual and to define his rights and freedoms should be applied
without regard to race or color.

Why, then, examine the question of average racial differences in genetic
endowment? The question remains important; first, because of efforts to place
individuals in jobs, schools, and housing on the basis of race without regard
for their abilities, drives, or behavioral standards; second, because of efforts
to extend the concept of equality among races to a point where it conflicts
with the rights of individuals to move ahead according to drive and ability in a
free competitive society; and third, because of a growing body of opinion that
the way to solve racial conflict is by the interbreeding of races. If it is true that
there are significant differences between the genetic endowments of the races,
this knowledge could and should affect the handling of biological and sociological problems.

If it were possible to prove the existence of significant differences between
representative samples of white and Negro brains in respect to any biological
measure which could be correlated with intelligence and behavior, this would
be evidence for a biological basis for racial differences.   Carleton Putnam
teaches (2) that such evidence exists and that for this and other reasons Negroes
should be segregated socially and in schools. The evidence cited by Putnam is
without scientific value. He has referred to Bean (3), Vint (4), and Connolly (5)
as having described morphological differences in the brains, especially the frontal
lobes, of Negroes and whites. There is no objective evidence that any of the
average differences claimed to exist are marks of inferiority or are correlated
with intelligence and behavior. I have reviewed the studies by Bean, Vint, and
Connolly in detail (6) and found that none of the relevant variables were controlled in any of the studies. The brains were from cadavers. Vint studied

only Negro brains and had no control group, and Bean and Connolly compared Negro brains from American sources with brains obtained decades ago from Germany. It would be difficult to imagine more inadequate sampling. Putnam referred to Herrick (7), Halstead (8), and Penfield and Rasmussen (9) on the importance of the frontal lobes. Nothing is said by these authors which would support Putnam's claim that the Negro brain is inferior; in fact, Halstead, Penfield, and Rasmussen have repudiated the teachings of Putnam (10).

How should the claim that the Negro brain is genetically inferior be tested? It would be necessary to gather brains from persons of different races for whom the respective environments, including prenatal and postnatal nutrition, were equivalent. The factors of age, sex, and health would have to be controlled. Diseased and senile brains should be excluded. The brains would have to be removed from skulls at the same time after death, fixed, processed, and measured by identical methods, and studied as "unknown" by not one but several experts. If it were possible to establish significant average differences in brains among races, it would still remain to be shown that any such difference is a mark of inferiority or that it is a physical measure of intellect or any other quality of mind. Claims to knowledge must also withstand replication of the experiments as well as debate and criticism.

It is possible that differences in the structure and chemistry of brains do have a relation to differences in race, but at present there is no satisfactory evidence for or against this proposition.

What kind of evidence would be needed to settle the question of race and intelligence? First, we need valid, culture-free measures of intelligence; second, representative samples of white and Negro populations as subjects; third, statistical determination of the significance of the differences in test performance; fourth, replication of the studies by different designs and methods; and finally, debate and criticism of methods and results in order to exclude other plausible hypotheses. Since only culture-linked tests of intelligence are available, scientists aim to compare representative samples of whites and Negroes in which the culture (home, street, and school), socio-economic status, and other relevant variables are the same. Now we reach a basic problem: Performance on intelligence tests, culture, and socio-economic standing are all correlated, because cultural and socio-economic advancement depend in part upon intelligence; a sample which is atypical with respect to one factor will be atypical with respect to the other two variables. This consideration is damaging to the claims of the equalitarians.

Genetic Basis for Racial Differences

What evidence is relevant to the claim for a genetic basis of racial differences in drives and intelligence? All of it is indirect. (i) Studies of experimental animals show that ability to learn has a genetic basis and that drives and other behavioral traits have an important genetic component *(11)*. (ii) Studies on man have shown beyond reasonable doubt that ability to learn and reason has a genetic basis. The most convincing studies are on identical twins *(12)*; the same studies also demonstrate the importance of environment. (iii) The histories of the Negro and white races show that the latter have made greater contributions to discovery and social evolution. (iv) Negroes on the whole perform less well in school and on objective intelligence and achievement tests. This is generally true even when the Negro child has attended integrated schools from the beginning. The culture of the home and street probably remains unequal in most, if not all, cases. There is a general tendency for the difference between the white and Negro children to increase as they grow older. The equalitarians argue that the average decline in IQ scores of Negroes results from substandard schooling and increasing experience with social discrimination. An alternative hypothesis is that this decline is due in part to a genetically based difference in maturation of ability to learn and manipulate abstract ideas. (v) With some welcome exceptions, Negroes tend to do poorly in the sciences and medicine. This is sometimes true of Negro students who have experience only with good home environment and good schools. This consideration, not far advanced beyond case-study evidence, is not compelling. (vi) It seems improbable that when races differ in other physical characteristics, the human brain, the highest product of evolution, would show an identical distribution of capacities among the races.

What considerations favor the equalitarian view? (i) Tests of intelligence and school achievement are affected by culture, and each can be improved by coaching *(13)*. Intensive training and motivation of Negro children has sometimes raised their average test performance to equal or exceed norms for whites. Missing from such studies are control groups of whites given extra training and motivation. Changes in culture, motivation, and coaching improve the achievement and test performance of white as well as Negro children. A significant improvement in the average test performance of recruits between World War I and World War II is reported *(14)*, both whites and Negroes showing gains, so that a marked difference between the races remained. It is not known whether optimal support of genetic endowment of both races would decrease, abolish, or increase the differences between races.

(ii) In a few, though not in all, studies in which Negro and white children were matched for socio-economic background, the differences in school achievement and test performance were insignificant *(15)*. As I mentioned above, this evidence is faulty, since intelligence, culture, and socio-economic standing are all correlated. It could be argued that because the Negro is handicapped by job discrimination, he must have higher intelligence than the white in order to reach the same socio-economic level. Some other investigators report that, even when whites and Negroes are equated for socio-economic standing, the school and test performance of the whites averages higher than that of the Negroes *(16)*. It may be that the groups are only superficially equated and that they remain different in several important respects.

(iii) The test performance of Negro recruits from certain northern states in World War I was superior to that of white recruits from certain southern states (17). Except to illustrate the overlapping of achievement and performance of the races it is absurd to omit comparisons between northern whites and northern Negroes and between southern whites and southern Negroes and conclude that racial differences are due to environment only. The cornerstone of the scientific method is the experiment in which all relevant variables except the one being studied are controlled. It is claimed that the average difference in test performance between white and Negro recruits was greater in World War II than in World War I (18).

(iv) Some studies of learning in infants and young children have shown (19) that the Negro child performs as well as or better than the white child. The validity of the measure, sometimes simple motor learning, as a criterion of intelligence is the issue here.

(v) There is a growing body of evidence that the mother-child relationship in primates (20) and the early handling of laboratory animals (21) have important enduring effects upon biology and behavior. Strodtbeck (22) has studied the factors which initiate a syndrome of poor socialization in the children of Negro slums and has shown that special training can facilitate the readiness of these children for school.

The above references are given to illustrate some relevant studies and problems. Most of the evidence on racial differences is descriptive and gives little insight into the causes of these differences beyond indicating that the pattern of causes is complex. Anastasi (23) has written an excellent treatise on differential psychology in which she emphasizes the possible environmental causes of racial differences. After reviewing some of the physical traits relating to race, this author states, "It appears improbable that racial differentiation in such physical traits was accompanied by differentiation with regard to genes affecting intellectual or personality development." Shuey (24) has covered the same sort of evidence but has emphasized the genetic basis of racial differences in intelligence.

The concept that the white and Negro races are approximately equally endowed with intelligence and drives remains a plausible hypothesis for which there is faulty evidence. The concept that the average Negro is significantly less intelligent than the average white is also a plausible hypothesis for which there is faulty evidence. There is no sound structure of evidence and logic which compels a conclusion on the issue of race and intelligence.

Sources of Data

How can new useful data on racial differences in abilities be obtained? I suggest the following possible sources of information. (i) The Armed Services of the United States have data on the test performance of large samples of recruits, extending through several decades of social progress for the Negro, that have not been fully reported. (ii) The school systems of some small northern cites have never segregated the races and have or could obtain comparisons of racial groups in respect to test performance and achievement. A number of school systems that have been desegregated for several years have new data on white and Negro children who have had experience only with good equal schooling. Some school systems do not record the race of the child, thereby making information on important issues unobtainable. It may, however, be impossible to find culturally equated samples of whites and Negroes in Anglo-American societies. (iii) Similarly, there are communities, especially in western states and Canada, which have never practiced segregation other than social (which may be important) or job discrimination. (iv) There are some orphanages in which children of different races have been housed and educated together from early life. (v) Finally, I suggest comparisons of the highest achievers of different races who have never experienced either a substandard culture or poor schools. Although the problem of sampling and unequal culture could not be completely controlled, I believe that, if sampling were done according to the best guidance by statisticians and if the problem were approached by different designs and methods, it would be possible to significantly reduce the faults in the evidence on race and intelligence.

Assurance of full civil rights for the Negro is desirable and necessary, but it will not solve all of his problems. To achieve the ideal that individuals be judged according to abilities, drives, and behavioral standards without regard to race will ease but not solve the problems of Negroes or members of other groups who are handicapped by substandard culture and possibly by poor genetic endowment.

The more militant Negro leaders who now dominate the civil rights movement, having been told that there are no genetically determined racial differences in drives and abilities, are demanding equal representation in jobs and in government at all levels of competence. If the equalitarian view is correct, then this is a just goal for the Negro. If it is not correct, then equal representation of the Negro at the higher levels of job competence and in government will be deleterious to society; return to the principle of judging the employability of the individual without regard to race would be preferred. If the 30 percent overlap usually found between the test scores of whites and Negroes in the United States should indicate the extent of a genetic difference between the races, this means that the number of Negroes who will remain dependent on social welfare is substantially greater than the number of whites so maintained and that the number of Negroes of high competence is substantially smaller than the number of such whites.

I have no doubt that forced segregation of the Negro in schools has generally had a deleterious effect upon the Negro child. I believe that voluntary integration of schools based upon compatible abilities, drives, and behavioral standards is wise and just. When schools are desegregated as a result of social pressures or when desegregation is forced without regard for abilities and behavioral standards, then the standards of the school are downgraded; those Negro children who are unable to compete at a high level are placed in special classes, and the school eventually becomes resegregated. Many individuals believe that forced random desegregation of schools and housing will solve the problems of the Negro. I oppose equally both forced segregation and forced desegregation in schools and housing; both are affronts to individual freedom and private judgment.

The aim of some integrationists is identical education for all without regard to drives and abilities. Achievement of this aim would be a blow to those public schools which have made some progress toward meeting the differing needs of the dull, the average, the bright, and the gifted of all races.

There are many competent citizens among Negroes and other racial groups who need better housing and who make good neighbors. Increased construction of middle-class housing open to all racial groups would permit voluntary integration without fomenting racial hate. A number of successful examples have already been developed by private enterprise.

The idea that forced random desegregation meets opposition solely because of racial prejudice is pure fiction, a grave error of judgment against many individuals who fear desegregation. Racial prejudice is real, ugly, and powerful, but its causes are such that it cannot voluntarily be set aside upon command or persuasion. There are compelling reasons why the average white does not wish to have the average Negro as a neighbor or schoolmate which have nothing to do with the color of skin. I refer to poor behavioral standards, none of which are uniquely Negro. For similar reasons Negroes of high standards seek to escape bad culture and would not choose to live with members of any race who represent it. These are the people, not all of them Negro, who are buffeted about by social pressure, who are removed and largely forgotten by urban renewal projects; they are unwanted. These problems cannot be solved by forced integration.

These are facets of the larger problems of overpopulation, crime, unemployment, and the rising costs of relief which will surely be increased by automation. Most attacks upon these problems are merely palliative; there are some successful attempts to salvage damaged lives, but there have been no major efforts to prevent these social and biosocial problems. I believe that we have the basic knowledge needed to prevent most of them. We should discuss and debate possible methods, although society is not ready to adopt them.

Improving Substandard Culture

The handicap of substandard culture begins to be transmitted from adult to child in the earliest formative years. If the vicious cycle of culturally handicapped adults–culturally handicapped children–culturally handicapped adults is to be interrupted, this should be undertaken at birth. Ideally, slum clearance should be undertaken on a giant scale all over the nation. Poverty is pathogenic to many individuals. The areas cleared would be replaced by subsidized apartment buildings and family houses, some but not all communities being desegregated. An intensive program of conception control should be established; the underprivileged are commonly denied information on conception control, although such information is readily available to those outside the slums. Conception control is important for all who, either because of genetic limitations or because of poor cultural heritage, are unable to endow children with a reasonable chance to achieve happiness, self-sufficiency, and good citizenship. The very high birth rate among indolent incompetent Negroes is a threat to the future success of this race. I do not suggest any programs which would threaten the genetic diversity of man; the same standards of genetic offensive would be applied to all races. Conception control and eugenics, especially the increased use of artificial insemination (25, 26), are means of achieving gains in genetically determined competence, and no class or race should be denied their use as a means to biological and social evolution and prevention of human misery.

Equally important would be nursery care and youth programs which would place each child in an environment favorable to his making the most of his native abilities. Although my statement of the problem here and of possible corrective measures is simplistic, such programs are not impossible. The Scandinavian countries, Denmark especially, have almost eliminated slums, and the incidence of crime is low compared to America. As recently as the 1930's, the Soviet Union had gangs of wild children who lived by robbery and violence. By intensive nursery care and youth programs, the USSR claims to have reduced delinquency to a low level. Of course, a primary purpose of youth programs in the collective societies is to indoctrinate children into their political philosophy. We need not emulate their objectives or all of their methods; surely, there is an American way to occupy the time of infants, children, and young people in a way that will teach good citizenship. The 4-H program is as American as corn-on-the-cob. Although designed for farm boys and girls, it has been tried successfully in urban areas. Extension of the 4-H program into all parts of the country and development of it to embrace science projects and other kinds of knowledge and skills might be one important step toward preventing crime and developing self-sufficiency and good citizenship. A number of important youth organizations exist, but most of them are for "good" boys and girls and are out of touch with the underprivileged.

The cost of subsidized good housing for all who need aid would be enormous. It has been estimated that in the absence of special effort there will be by 1970 about 3.2 million dilapidated units and 5 million units lacking one or more plumbing facilities (27). On the assumptions that the unit cost of replacement is $15,000 and that half of the dilapidated units must be replaced, the cost of the 1.6 million replacement units would be $24 billion. Arbitrarily allowing one-fourth as much per unit to restore the other 1.6 million dilapidated units would cost $6 billion. Finally, if $1000 per unit for modernization of the 5 million substandard units is allowed, this would cost an additional $5 billion.

These are conservative estimates. It is easy to imagine that the complete replacement of all substandard housing with good housing would cost $50 to $60 billion or more. The added costs of subsidy to insure a decent standard of living and extensive nursery care and of child and youth programs would add billions of dollars per year.

Cost of Improvement

What does bad environment cost America? It is estimated that the direct and indirect costs of crime exceed $30 billion per year. I do not know the reliability of such estimates. The city, state, and federal costs of relief are large and are growing. The cost in human misery is inestimable. The problem is complex. Not all of the underprivileged have a substandard culture or become dependent or criminal; the middle and upper classes of society also contribute to dependency and crime.

Considering the grave dangers of overpopulation, an intensive conception control program among those who for either cultural or biological reasons are unqualified for parenthood would be far cheaper and effective. The procedure for sterilization of each sex is now simple, and there are mechanical devices for preventing conception in the female that can remain in place for years without attention or apparent harm. I suggest that barrenness could be economically subsidized. Although society accepts a number of restrictions on mating, a program to prevent conception among those who are biologically and culturally unsuited for parenthood will be opposed on legal, ethical, and religious grounds.

Relatively few human geneticists claim that knowledge of human inheritance is sufficiently advanced to plan wisely a program of eugenics. Its use would involve risks in a world where governments achieve greater and greater control of personal freedoms without accompanying advances in ethics and wisdom. Some of the programs that have been attempted or proposed in the past are not reassuring. But I believe that we must take the risks and work toward the linking of euthenics and eugenics. The proposals of Muller (25) and of Huxley (26) represent sound biology and do not threaten the genetic diversity of man. Programs to control conception are needed now. It is not too early to expand educational and advisory programs; we can teach the use of artificial insemination when the husband but not the wife is a known carrier of genetic defects. With gradual evolution of eugenics, society may be ready to use the knowledge which may come several decades hence of how to control human heredity by changing the germ plasm.

Those who hope for the equality of all men without thought of their biology should be asked, "Shall we aim to make all men sick or all men well? Shall we aim toward universal incompetence or universal competence?" The concept of equality is meaningful only as it relates to civil rights and opportunities. Otherwise, to aim for the complete equality of all men is an affront to basic freedoms and rights of each individual to seek self-fulfillment according to his interests, drives, and abilities.   The ideal of letting each individual move ahead in a competitive society according to his drives and abilities will be realized only if the individual is biologically fit for competition and is free from the almost insurmountable handicap of slum environment.   This aim suits the United States, which intends to remain a free competitive society.   The philosophy which abhors competition and holds that men are born biologically equal or, if they are not, that they should be kept equal by Procrustean methods, would establish mediocrity, not excellence, as a national goal.   It is a philosophy which exists to the left of red.

Although there is no possibility that a comprehensive program to upgrade the genetic and cultural heritage of all the races will be undertaken for several decades, it is not too early to make a beginning, especially to seek more complete information on the causes of biosocial problems, to propose and debate methods, and to test some ideas by pilot studies.   The guiding principle should be prevention of biosocial problems rather than to depend upon palliative methods.   Private enterprise should have a major role in defining goals and development of plans.   But both private and public enterprise need strong leadership from the Office of the President and other representatives of the people.

REFERENCES

1. W. C. HALSTEAD, *J. Psychol.* **20,** 57 (1945).

2. C. PUTNAM, *Race and Reason* (Public Affairs Press, Washington, D.C., 1961); "These are the guilty" [address before the Washington Putnam Letters Club, 12 Feb. 1963. Reprinted in *Mankind Quart.* **4,** 28 (1963)].

3. R. B. BEAN, *Am. J. Anat.* **5,** 353 (1906).

4. F. W. VINT, *ibid.* **68,** 216 (1934).

5. C. J. CONNOLLY, *External Morphology of the Primate Brain* (Thomas, Indianapolis, 1950).

6. D. J. INGLE, "Comments on the teachings of Carleton Putnam," *Mankind Quart.* **4,** 28 (1963).

7. C. J. HERRICK, *The Evolution of Human Nature* (Univ. of Texas Press, Austin, 1956).

8. W. C. HALSTEAD, *Brains and Intelligence* (Univ. of Chicago Press, Chicago, 1947).

9. W. PENFIELD and T. RASMUSSEN, *The Cerebral Cortex of Man* (Macmillan, New York, 1957).

10. Letters to and from the editor, *Perspectives Biol. Med.* **6,** 539 (1963).

11. C. S. HALL, "The genetics of behavior," in *Handbook of Experimental Psychology,* S. S. STEVENS, Ed. (Wiley, New York, 1951).

12. H. H. NEWMAN, F. N. FREEMAN, K. J. HOLZINGER, *Twins: A Study of Heredity and Environment* (Univ. of Chicago Press, Chicago, 1937).

13. W. S. JAMES, *Brit. J. Psychol.* **23,** 155 (1953).

14. R. D. TUDDENHAM, *Am. Psychol.* **3**, 54 (1948).

15. F. BROWN, *J. Genet. Psychol.* **65**, 161 (1944).

16. F. C. J. McGURK, *J. Abnorm. Soc. Psychol.* **48**, 448 (1953).

17. O. KLINEBERG, *Characteristics of the American Negro* (Harper, New York, 1944).

18. J. B. MINOR, *Intelligence in the United States* (Springer, New York, 1957).

19. A. R. GILLILAND, *Child Develop.* **22**, 271 (1951).

20. H. F. HARLOW and M. HARLOW, *Sci. Am.* **207**, 136 (1962).

21. S. LEVINE, *Acta Endocrinol. Suppl.* **51**, 41 (1960).

22. F. L. STRODTBECK, unpublished manuscript.

23. A. ANASTASI, *Differential Psychology* (Macmillan, New York, 1961).

24. A. M. SHUEY, *Testing of Negro Intelligence* (J. P. Bell, Lynchburg, Va., 1958).

25. H. J. MULLER, *Perspectives Biol. Med.* **3**, 1 (1959).

26. J. HUXLEY, *ibid.* **6**, 155 (1963).

27. R. C. WEAVER, *16th Annual Report, Housing and Home Finance Agency* (U.S. Government Printing Office, Washington, D.C., 1963.)

Now that you have finished the article and formed your own independent judgments (but, then, just how independent can your judgment be from the past influences and prejudices to which you have been exposed), proceed to read the letters to the editor* written in response to Dr. Ingle's article.

Confusion of Issues

It is not clear why standards of conceptual and definitional precision should not be as rigorous for essays labeled "heuristic" as we expect them to be for more conventional scientific reports based on research and analysis. Aside, for example, from the questionable scientific utility of using anything at all "in its popular sense," one might fairly inquire in *which* popular sense Dwight Ingle ("Racial differences and the future," 16 Oct., p. 375) uses "race," among other terms. Presumably he also uses "racists" and "equalitarians" in some popular sense.... Not all "racists" do in fact maintain that Negroes are genetically inferior. Nor do all "equalitarians" maintain that "all races are equally endowed with intelligence." Such is the case, at least, if "equalitarian" is extendable to the scholar who, *qua* scientist, could not offer such a contention in the absence of reasonable proof. The point is (can Ingle be unaware of it?) that in the absence of firm evidence to the contrary there is no justification for assumptions that racial groups are differentially equipped in terms of such potential as is indeed genetic for intellectual, cultural, and emotional development.

---

* *Science*, 11 December and 18 December, 1964. Reprinted by permission.

There is, then, no binary opposition between *the* "racist" and *the* "equalitarian" position. There is, rather, a large variety of positions, not all of which are conceptually discrete. Most scholars who are trained in human biology, genetics, psychology, anthropology, and so on are, *qua* scientists, neutral (an alternative not offered by Ingle's dichotomy) on ethno-racial issues, since confirmatory evidence is lacking.

They maintain *qua* citizens, however, that there is no scientifically justifiable reason to deny racial groups per se access to those opportunities and privileges that our Constitution guarantees all citizens. Ingle does not mention that an international committee of biological scientists under the auspices of UNESCO (1952) and the American Anthropological Association (1961) have made the same points in formal statements.

I doubt that Ingle's simplistic reduction of reality to two mutually exclusive and opposed categories has even the "heuristic" value that he claims for his thoughts. Nor does such phraseology as "the average Negro" and "the average white" impress me as very useful for comparing groups on multifactor traits. . . .

The statements by the UNESCO group and the American Anthropological Association explicitly and rightly contend that there must be a separation of issues: that the scientific problem of possibly significant biopsychological differences between racial groups and access by members of such groups to Constitutional guarantees are separate questions, not to be confused. Ingle confuses them in several places. For example, he argues that if it can be shown that there are "genetically determined racial differences in drives and abilities," then "equal representation of the Negro at the higher levels of job competence and in government will be deleterious to society." The implication, presumably, is that under certain circumstances the fundamental guarantees of the Constitution can and should be abridged.

Ingle will have his own reasons for his confusing of these issues, just as he will have his reasons for believing that "voluntary integration of schools [as opposed to legally enforced integration] . . . is wise and just." The question is: Of what heuristic value are such expressions in the pages of *Science?*

JAMES RICHARD JAQUITH

Department of Sociology-
Anthropology, Washington Univ.
St. Louis, Missouri

Problems of Our Own Making

Ingle writes, "Racists claim that the Negro race is genetically inferior to other races in intelligence, while equalitarians claim that all races are equally endowed with intelligence." Since, I take it, "equalitarians" means those to whom that label has been most frequently applied, namely, Boas, Klineberg,

Herskovits, myself, and the majority of scientists who have written on the subject, it is in this particular connection very necessary to correct the misstatement about the "claims" of equalitarians. It should be clearly understood that the term "equalitarian" is one customarily used by racists to describe those who are critical of their views. This is a typical racist device: to distort and misrepresent what their critics have in fact stated, and then to label them with a term which further distorts the views. . . . What should be implied by the term "equalitarian" is the belief that every human being has an equal birthright, which is development. In this sense all men of good will are, I hope, "equalitarians."

It is possible that, to use Ingle's highly inaccurate phrase, "all races are equally endowed with intelligence," but until the great experiment has been performed of allowing the members of all groups called "races" equal opportunities for development we shall never know whether they are or not. No group, "race," or individual is endowed with intelligence. Individuals are endowed with genetic potentials for learning to be intelligent. Intelligence is a socially acquired ability, a complex problem-solving form of behavior which one must learn from other human beings. Not only that, human beings have to learn to learn. The capacity for intelligence becomes an ability only when it has been trained. The capacity itself varies among individuals and, allowing for differences in prenatal influences, these capacities are largely genetically influenced. Allowing for the genetic differences, all observers are agreed that what those capacities will become as abilities will largely depend upon the environmental stimulations to which they are exposed. . . .

Ingle informs us that "The histories of the Negro and white races show that the latter have made greater contributions to discovery and social evolution." By this, I take it, he means that whites have made greater contributions to discovery and social evolution since the Neolithic or the first and second industrial revolutions, say roughly within the last 12,000 years? For before that time, all men were living at a food-gathering-hunting stage of cultural development. Since we know practically nothing of the pre-history or archeology of the Negroid peoples before the Neolithic, it is not possible for anyone to say what contributions to discovery and social evolution they may or may not have made. But since Africa is agreed by most authorities to have been the original homeland in which the greater part of man's evolution, both physical and cultural, occurred in the prehistoric period, it is probable that some, if not all, of these people were Negroid, and that they made fundamental contributions to discovery and social evolution. With respect to more recent history it may be true that white peoples have made greater contributions. Achievements imply opportunities, and now that some African peoples are being increasingly provided with opportunities we may not have too long to wait before the returns start coming in. If it takes a hundred years, I should consider it, by the measure of the rate at which these changes have occurred in the past, very rapid indeed.

Ingle writes, "It seems improbable that when races differ in other physical characteristics, the human brain, the highest product of evolution, would show an identical distribution of capacities among the races." ... It should be clearly understood that the gene differences relating to the physical traits characterizing "races" are of very small number, and in any event, to reason from the existence of superficial, adaptive physical differences to the existence of significant behavioral differences is to misunderstand the nature of the conditions and modalities involved.

"The more militant Negro leaders," Ingle writes, "who now dominate the civil rights movement, having been told that there are no genetically determined racial differences in drives and abilities, are demanding equal representation in jobs and in government at all levels of competence." The American Negro's struggle for his elementary rights is not based on what "equalitarians" or anyone else may have told him, but upon the irrepressible drive and the inalienable natural right of every human being to enjoy the satisfaction of his needs for development. This has nothing whatever to do with what anyone may have told his "militant" leaders.

Ingle says he has no doubt that "forced segregation of the Negro in schools has generally had a deleterious effect upon the Negro child." Nevertheless, he writes, "Personally, I oppose equally both forced segregation and forced desegregation in schools and housing; both are affronts to individual freedom and private judgment." "There are compelling reasons," he writes, "why the average white does not wish to have the average Negro as a neighbor or schoolmate which have nothing to do with the color of skin. I refer to poor behavioral standards, none of which are uniquely Negro." The fact is that the "poor behavioral standards" of many Negroes were created by and are the direct result of the treatment Negroes have received from the white man. Surely it would be a small thing to ask of whites, in repayment of the enormous debt they have accumulated for their past crimes against the Negro, to help him raise his "standards," at whatever level they may be. And it might just happen that some whites would find themselves learning from their Negro neighbors that there is more to being human than the proper "behavioral standards."

"The very high birth rate among indolent incompetent Negroes is a threat to the future success of this race," writes Ingle. I don't see that this is any more true of Negroes than it is of any other indolent incompetent individuals and their effects upon the "race." I wholly agree with Ingle that "Conception control is important for all who, either because of genetic limitations or because of poor cultural heritage, are unable to endow children with a reasonable chance to achieve happiness, self-sufficiency, and good citizenship." I also entirely agree that "The guiding principle should be prevention of [social] problems rather then to depend upon palliative methods." But the chronically irritating fact is that we are squarely faced with considerable social problems of our own making, which we did nothing to prevent. It is not too late to do many things directed at preventing their further exacerbation. The teaching of birth control is an imperative, and so is the institution of other social means by which we might achieve the solution of many of our social problems. In this connection such genetic differences as may exist between American Negroes and American whites are of no relevance whatever.

Let us work toward the development of a society in which everyone is afforded the opportunity for self-development, and then let us observe what happens. This seems to me the only practical approach to the problem of human relations in any society. In spite of very real appearances to the contrary, I believe that this is the direction in which humanity is traveling.

ASHLEY MONTAGU

321 Cherry Hill Road,
Princeton, New Jersey

Whose Bad Culture?

... We are familiar with the eugenics argument that individuals or classes or races should be prevented from reproducing their kind because of an alleged genetic defect. What I find novel and mischievous in Ingle's discussion is the recommendation that adherence to a culture that is definable, somehow, as "bad" or "substandard" (Ingle's terms) should be taken as grounds for eugenic measures ("conception control"). Ingle directs our attention to the effect on the American Negro of the heritage of slavery, and the difficulty that successive generations experienced, into the present, in breaking out of the "vicious cycle of culturally handicapped adults—culturally handicapped children—culturally handicapped adults." Here he seems to face the problem squarely as a social one and suggests for our consideration a pair of measures that ... might have a meliorative effect on this self-compounding process: a gigantic slum-clearance project, and "nursery care and youth programs which would place each child in an environment favorable to his making the most of his native abilities." He compares the cost of such a program with the cost of crime and of relief. Well and good as far as that goes ... But now follows a strange turning in his presentation. Having outlined a partial program of environmental improvement, he demurs abruptly to the efficacy of these undertakings ("The problem is complex ..."). He then goes on immediately to propose another course of action, either as replacement or as accompaniment to the program of environmental melioration, it is not clear which. "Considering the grave dangers of overpopulation," he says, "an intensive conception control program among those who for either cultural or biological reasons are unqualified for parenthood would be far cheaper and effective." ... Having placed a large question mark against the two programs of environmental improvement that he outlined, he proposes one policy (conception control based on biological insufficiency) that is not justified by his discussion of the biology of race, and a companion policy (conception control based on cultural insufficiency) whose ethical ruthlessness overshadows its sociological ineptness. ...

The judgment that this proposal is ruthless is a matter of taste and perhaps not arguable. But I would like nevertheless to have us reflect on one aspect of

Ingle's proposal, its effect of continuing a historical injustice. . . . The Negro's condition as a slave forced him into the position of a backyard adjunct to white society. The rudimentary social right of maintaining a family was denied him in many cases. . . . His very color and physiognomy have become symbols of inferiority, functioning to confirm him in his place, blocking achievement, or, where achievement has occurred, over-shadowing it. Ingle's proposal would be effective, as he claims; but death is also effective as a solver of earthly problems. It would be "cheaper," too, as he claims. (It is not clear whether he means cheaper than meliorative housing and youth programs or cheaper than the costs of crime and relief to which he refers. But no matter; it would be cheaper.) . . . When all this is said, we must ask whether we could really opt for such an easy way out of the consequences of our actions toward the Negro. . . .

After stating his radical proposal for conception control, Ingle advances in the very next sentence to tell us that "the procedure for sterilization of each sex is now simple," and in the following sentence suggests that "barrenness could be economically subsidized." There is an unseemly haste in this sequence of technical observations. Not even the looming problem of overpopulation seems to justify such haste in implementing a program that is so questionable in its foundation. For a question that is central to Ingle's proposal remains unanswered (I think it is not even raised): by whose standard is anyone's culture to be judged as a disqualification for parenthood? Each of us privately maintains his own judgment as to what is bad culture and bad behavior. I will be so bold as to say that I consider Ingle to be manifesting bad culture himself when he launches forth on this awesome set of proposals in the detached spirit of a sanitary engineer. But I am far from advocating the sterilization of him and his kind.

A psychiatrist, commenting on a similar problem of classification in his field, wrote, "It has been said, with a good deal of truth, that what a man with a character disorder has is a 'character that the diagnosing individual disapproves of' " [Martin Hoffman, *Yale Review* **54**, 22 (1964)]. Can we presently, or foreseeably, hope for markedly better results than this in diagnosing cultural disorder? . . . The very indolence with which the Negro is charged, which is so unacceptable to true upholders of the Protestant ethic, is passing into the majority society in a form that I would call spiritually enhancing and psychologically beneficial. The pattern of nonviolent resistance developed by Southern Negroes in recent years is in one view feckless, impertinent, and disruptive, and in another a heartening improvement over our traditional reliance on violent methods of struggle. . . .

The scientist should be encouraged to explore where his fancy carries him, but, when it comes to stating social policy, it is extremely irresponsible to base recommendations on an unproven hypothesis, and to slight a rich field of action that bases itself on demonstrable relations between a massive historical cause and a widespread social defect. . . . It would be humane and sociologically sound to try with all the resources at our command to introduce into the flow of contemporary history such actions as will help Negroes to direct themselves toward a more rewarding existence—resources, incidentally, which the Negro helped open to us by several centuries of unpaid and underpaid labor. . . .

JULES RABIN

14 Bedford Street, New York City

Political Physiology

. . . Possibly Ingle may be forgiven for knowing far less about politics than about physiology. . . . He should know that the demands made by civil rights groups have never been for jobs for Negroes regardless of ability, but for jobs precisely on the basis of ability, without regard for color. There are demands for special training and opportunities for Negroes to make up for past injustices, but such demands have no connection with genetic differences or lack of difference. As for opportunity to enter schools or to obtain housing wherever vacancies exist, Ingle's unstated assumption is that genetic differences might serve as justification for maintaining segregation. Possibly he would like to have landlords give I.Q. tests to prospective tenants, but this would be a political question, having little to do with either genetics or physiology.

Ingle says that there are efforts to extend the concept of racial equality to the point where it conflicts with "the rights of each individual to seek self-fulfillment [and to] move ahead in a competitive society." The Negro who would like to become a salesman, a bank clerk, plumber, electrician, or brickmason notes that the only genetic difference of interest to the prospective employer (or labor union) is the color of his skin, and he has long found this an obstacle to self-fulfillment and to his right to move ahead competitively. Ingle refers to "poor behavioral standards" as one justification for opposing forced integration. It is inexcusable that he ignores the obvious fact that, no matter how "well-behaved," Negroes have difficulty obtaining housing in white communities.

The setting up of such straw men as "forced integration" and "the philosophy which abhors competition" is a standard tactic in political speeches, but has no place in a scientific paper. Possibly Ingle should study the genetics of such straw men. That would be more useful to science than is his unfortunate article.

ABRAHAM S. ENDLER

150-24 78 Avenue,
Flushing, New York

Trading upon Science

. . . As a "source of data" Ingle suggests "comparisons of the highest achievers of different races who have never experienced either substandard culture or poor schools." Difficulties spring to mind at once: (i) It might be difficult to find Negro "high achievers" who met the condition. Negro writer James

Baldwin, whose talents were nourished in the Harlem slums, would be excluded, for instance. (ii) What is a "highest achiever" anyway? A handicapped person who triumphs in a small way over his handicap? Earner of more than $100,000 a year? Nobel Prize winner? One who has fought to the top in the savagely competitive sport or entertainment world? Reformer? Pillar of the Establishment? Successful rogue? Professor of physiology? (iii) Suppose some comparison of achievements were possible. Would magnitude of accomplishment measure the intelligence of its agent? How to evaluate the difficulty of the achievement, obduracy of circumstances, the factor of luck, the cooperation of others? These are fairly obvious difficulties, suggested by common sense. A slight acquaintance with psychology suggests many others. . . .

Ingle suggests that those who emphasize environmental effects are misled by an "equalitarian dogma." Such a statement discounts work by Piaget, McCulloch, Pitts, Rosenblatt, and others on formal theories utilizing environmental feedback mechanisms. This work is not inspired by such a dogma; it arises from the assumption that internal cognitive structures are unknowns, to be discovered empirically. That environment is very important in the organization and modification of such structures is a theory which has proved to be scientifically fruitful in investigating learning and intelligence, not something that was assumed a priori.

Ingle's hypothesis that there may be a genetic basis of intelligence and that we ought therefore to upgrade our genetic heritage, or ought to breed for more intelligent people, is open to objections other than that we don't know what intelligence unrelated to culture is. W. R. Thompson and T. L. Fuller have shown that there is little if any relation between genotype and phenotype (roughly, heritage and character), at least for such traits as are polygenic [W. R. Thompson, *Eugenics Quart.* 4, 8 (1957)]. As for a genetic basis for cultures (or subcultures), also suggested in Ingle's article, the comments of Steward and Shimkin [in *Evolution and Man's Progress*, H. Hoagland and R. W. Burhoe, Eds. (Columbia Univ. Press, New York, 1962)] seem relevant:

A demonstration that genetic factors have shaped cultural patterns will require a rigorous scientific methodology that has not been developed. The assumption that individuals can be bred for superior culture not only lacks scientific validation of the relation between genetics and culture, but presupposes indefensible conclusions concerning the superiority of any culture.

Given the known difficulties, the thesis that a racial basis for intelligence can be tested seems not at all worth debate. A program to test now for a genetic basis of intelligence is a mischievous suggestion, not a scientific speculation.

A curiosity in the Ingle piece is worthy of note: Why should "private enterprise . . . have a major role in defining goals and development of plans" for upgrading "the genetic and cultural heritage of all the races" by sterilization, various economic subsidies, and Little Dandy Superior Sperm Banks for the genetically underprivileged?

And finally, a major curiosity: Why was the article published? Since publication in *Science* confers a cachet of some minimal scientific respectability, the editors must certainly be responsible for screening out those manuscripts which seek to trade upon the neutrality of science, and their authors' competence as scientists, in order to engage in special pleading. Since one must suppose (now having evidence) that scientists, like others, succumb occasionally to their prejudices, the principal blame for this irresponsible publication, during an election campaign when civil rights of Negroes are a major issue, must rest with the editors of *Science*.

PAULA GIESE

5654 South Drexel Avenue,
Chicago 37, Ill.

An Analogous Problem

Ingle has suggested that, until we find out whether or not there are any important differences between races as popularly defined, we should supplement our present treatment of one of these "races" with a bold program of euthenics and eugenics. Not only would such a program be more humane than Swift's classic proposals for the handling of overpopulation in Ireland, but it could also be applied to the Bigot problem.

Although there is no trait which is found exclusively in a single race, it is well known that Bigots possess a substandard culture. The following points are commonly accepted: (i) Bigots tend to form closely knit cultural groups and to mate almost exclusively among themselves. This racial homogeneity may limit their biological variability and encourage the preservation of harmful mutations among them. (ii) Bigots tend to confuse biologically, sociologically, and popularly defined races, an obvious indication of their lack of any capacity for abstract thought. (iii) Bigots are good citizens, and their efforts to preserve our nation from mystical-magical pollution and impurity are surely commendable. Regrettably, Bigots are not good neighbors or good schoolmates; evidently this is related to the fact that they can perform well only on intelligence tests of their own devising. (iv) The common beliefs that Bigot males tend to form Saturday-night alliances with members of other ethnic groups (meaning cultural or genetic groups or something like that), that they tend to solve their problems through outbursts of uncontrolled aggression, or that they fear competition have been shown to be false. All racial groups (a popular concept used here in its strict biological sense) possess these characteristics; they are simply much more common among Bigots.

Action on the Bigot problem (by which I mean sterilize the lot of them) requires only the following steps: (i) We must find a biological definition of race which conforms to the prejudices of the paranoid and the uneducated and is also acceptable to biologists and physical anthropologists. (ii) We must find out what genes human beings have and which ones are responsible for substandard culture. We do not, of course, have good evidence to prove that Bigot aggressiveness is genetic, but we do know that mice bite. In any case, the concept of biosocial means that all human behavioral traits are biological unless proven otherwise. (iii) We must work out some way of sterilizing millions of Bigots without offending them. Unfortunately, this is a social science problem and therefore outside the scope of a paper conceived in purely biological and rational terms.

ALAN R. BEALS

*Department of Anthropology, Stanford
Univ., Stanford, Calif.*

Disastrous Definition

A line in Wigner's Nobel laureate address states that "the specification of the explainable may have been the greatest discovery of physics so far" (*Science*, 4 Sept., p. 995). If we substitute "science" for "physics" in that statement and apply it to Ingle's "Racial differences and the future" (16 Oct., p. 375), we find that article sadly lacking in basic scientific orientation. An investigator does not have to know everything about the subject of his research before beginning, but it is a minimum essential that he sufficiently delineate the subject of his inquiry so that he knows when he is looking at it and not at something else.

Take "race." Ingle says, "I use the word 'race' in its popular sense, recognizing that all ethnic groups represent mixed origins and that there is no known physical or behavioral trait which is found exclusively in one 'race'." To use the word "race" in its popular sense in an article which pretends to be scientific is disastrous. And whose "popular" concept of race is it that Ingle uses? . . . The reports of Wagley, Harris, and their students have thrown light on the phenomenon of racial identity as a matter of sociocultural perception. In a study made in Brazil, Harris and his students elicited no less than 40 racial types when they showed sample drawings to Brazilians. . . . The tendency in the United States today to reduce the range of human variation to a very few allegedly polar types is something of a reversal from the days of slavery, when finer distinctions were drawn. Today's white-and-colored, or Caucasian-and-Negro, fits the social problems of the time; an earlier age distinguished quadroons and octoroons and even further, but that went with the economy of the slave market.

Harry S. Murphy claims that he, not James Meredith, was the man who broke the color bar at the University of Mississippi. That few people in our society will identify Murphy as a Negro is patent from his acceptance at that institution, but he says he is a Negro and apparently has the genealogical evidence to prove it. "Prove it," that is, by the implicit—and in the legal codes of some states explicit—rule of descent that says that anyone with any Negro ancestor is a Negro. There are good grounds for making the statement that by that standard just about everybody in the United States is a Negro. Marvin Harris has expressed this very well:

Genetically speaking, about the only thing any racist can be sure of is that he is a human being. It makes sense to inquire whether a given creature is a man or a chimpanzee, but from the point of view of genetics it is nonsense to ask whether a particular individual is a white *or* a Negro. To be a member of a biological race is to be a member of a population which exhibits a specified frequency of certain kinds of genes. Individuals do not exhibit frequencies of genes; individuals merely have the human complement of genes, a very large but unknown number, most of which are shared in common by all people. When a man says "I am white," all that he can mean scientifically is that he is a member of a population which has been found to have a high frequency of genes for light skin color, thin lips, heavy body hair, medium stature, etc. Since the population of which he is a member is necessarily a hybrid population—actually all human races are hybrid—there is no way to make certain that *he himself* does not owe a genetic endowment to other populations ... The archaeological and paleontological evidence quite clearly indicates that there has been gene flow between Europe and Africa for almost a million years.... All racial identity, scientifically speaking, is ambiguous. Wherever certainty is expressed on this subject, we can be confident that society has manufactured a social lie in order to help one of its segments take advantage of another. [*Patterns of Race in the Americas* (Walker, New York, 1964), p. 55]

Ingle asks himself, "What kind of evidence would be needed to settle the question of race and intelligence?" His own reply is, "First, we need valid, culture-free measures of intelligence; second, representative samples of white and Negro populations as subjects...." He goes on with additional desiderata, but the damage has already been done. A colleague of mine once wrote a paper on "The occurrence of references to Buddha on oracle bones of the Shang dynasty." The paper was as brief as it was brilliant: since the Shang dynasty antedated Buddha by over a thousand years, there were no references to him on the oracle bones. Some problems have no solutions because they do not exist in the form in which they have been stated. No problem about race will be solved when it is stated in the matrix of archetypes; it is time for the 19th century to come to a close in racial anthropology, even among the amateurs.

MORTON H. FRIED

Department of Anthropology,
Columbia Univ., New York City

Intelligence in Modern Life

... Since the subject of social science is human behavior, it is difficult for even a trained social scientist to take an objective attitude toward his data. I have yet to hear of any scientist of any race proposing that members of some *other* race have superior average innate mental ability to those of his own race. There is an even greater danger in letting scientists in other branches than the social sciences, however distinguished in their own disciplines, publish in a general scientific journal statements on an issue with such explosive social implications as the evaluation of race. This is that lay groups concerned with political action and uninterested in scientific investigation will exploit any statements they can find to buttress nonscientific prejudices and even to justify violence against members of the allegedly inferior group and their partisans. Ingle's article is clearly subject to such misuse. . . .

I fail to see wherein Ingle's suggestions for "new useful data" on racial differences would provide any major breakthrough on problems which decades of previous research have failed to solve. His first suggestion, for instance, to go back to examining more carefully the results of more Armed Services tests of recruits, is hardly likely to produce any conclusive measure of innate racial differences because of the impossibility of controlling sociocultural factors in these data. The most sophisticated statisticians in the world will not be able to get clear answers out of basically inadequate data. . . . His suggestion to compare the highest achievers of different races "who have never experienced a substandard culture" likewise fails to show attention to the systematic, pervasive effects of racial discrimination. *Any* Negro raised in the United States has experienced a "substandard culture" from his own point of view. If some protected "favorable" community could be found, one would hardly expect the "highest achievers" to come out of it, since high achievement usually demands free communication in the society at large. . . .

Ingle hints that racial mixture may have the undesirable effect of producing a single population with lower average intelligence than the higher of the originally separate parent groups. Both the view that racial mixture among men leads to increased vigor and intelligence and the opposite view have been advocated and supported in anecdotal fashion. . . .   There appears to be a rather strong correlation between the development of large civilizations and racial mixture over the period of recorded history. . . .

Ingle apparently assumes that the chief goal of eugenics should be maintaining and improving the level of innate intelligence. This is a common enough assumption but one which I would raise some cautions about. In the first place, it is not at all certain that civilization as compared with "savagery" demands a higher level of intelligence in the sense of being able to integrate greater quantities of information in making a decision. The very characteristic of civilization is that it provides prefabricated answers and shortcuts to many of life's problems. It presumably requires much more intelligence and attentiveness to stalk and kill an antelope with a spear than with a high-powered rifle with telescopic sights. Thanks to increased population and improved communications and educational institutions in civilization, the ingenuity of one intelligent man can be spread rapidly much farther than that of a bright Paleolithic hunter. In the second place, the diversity of individual roles increases with the growth of civilization. It would seem to be a more suitable goal of eugenics to insure a reasonably diverse array of innate abilities in the population. . . . In the third place, any hereditary trait when maximized by selection is likely to entail some unintended, undesirable consequences. . . . And if we assume that intelligence is an unmixed blessing we may wonder how then to account for the considerable individual variability in intelligence, apparently in part genetically based, in all known populations. This could suggest a balanced polymorphism with respect to intelligence.

I am inclined to believe that eventually some sort of racial difference in innate mental abilities will be demonstrable, as they are in other respects, but doubt that these will be of much social relevance. The major races appear to have attained essentially their present form in Paleolithic times. Special genetic traits favoring adaptation to one or another cultural and natural environment in the Old Stone Age may be of little relevance to modern life.

J. L. FISCHER

Department of Sociology and Anthropology,
Tulane Univ., New Orleans, Louisiana

Methodological Note

Ingle's article is a very welcome contribution in terms of attitude, but by no means exhaustive in terms of possibilities. One of the most serious deficiencies in such studies and considerations is that, while the American white population represents a gene pool that is almost completely free of Negro influence, the American Negro gene pool cannot be said to be free, to the same extent, of white influence. This situation results from the fact that the two groups are defined socially rather than biologically. . . . Mulattoes are classed as Negroes in our society (on school records, for example), in spite of the fact that it makes as much sense to classify them as whites. Therefore, any comparison of American whites to American Negroes, especially a genetic comparison, is invalid to the extent that one is also comparing American whites to themselves.

This peculiar social situation may, however, provide in part the elements to resolve the question of genetic differences. One might compare the performance of "biological Negroes" with "social Negroes," since most of their environmental influences are the same.

NEIL B. TODD

Animal Research Center,
Harvard Medical School, Boston

Review of the Evidence

. . . Ingle presents less than an impartial view of the evidence. He finds space to refer to an unpublished manuscript, but does not mention the authoritative and careful review presented by Dreger and Miller [*Psychol. Bull.* **57,** 361 (1960)]. He submits to the reader Shuey's conclusion, although Williams

[*Contemp. Psychol.* **5,** 196 (1960)] characterized Shuey's book as "an exhibit of the futility of confusing summaries of data with the critical evaluation of what the data are supposed to represent. . . ."

DAVID A. PARTON

Pittsburgh Child Guidance Research Center,
Pittsburgh, Penn.

. . . Ingle states that "Studies on man have shown reasonable doubt that ability to learn and reason has a genetic basis" and cites Newell, Freeman, and Holzinger. But McNemar's critical review of that study concluded that "the only evidence which approaches decisiveness is that for separated twins, and this rests ultimately upon the fact that *four* pairs reared in really different environments were undoubtedly different in intelligence. This fact can neither be ignored by the naturite nor deemed crucial by the nurturite" [*Psychol. Bull.* **35,** 237 (1938)]. A far more recent and comprehensive review of twin studies, as well as other evidence, has been provided by Hunt, who concluded that "the assumptions that intelligence is fixed and that its development is predetermined by the genes is no longer tenable" [*Intelligence and Experience* (Ronald Press, New York, 1961), p. 362]. Furthermore, no unitary "ability to learn" has ever been satisfactorily demonstrated, and reasoning isn't quite so unidimensional either [J. P. Guilford, *Personality* (McGraw-Hill, New York, 1959)]. . . .

RICHARD E. SNOW
WARREN F. SEIBERT

Purdue Univ., Lafayette, Indiana

A Social Program

The most outstanding aspect of Ingle's article was the courage he displayed in his handling of what is certainly a hot potato. In addition, he made a significant contribution to the question of possible intellectual differences among the races by his public statement that not enough evidence exists to warrant a stand on the issue. I wonder, however, if there is as pressing a need as he suggests to find an answer to this question. I think rather that we should concentrate on ensuring that the Negro receives, as the author maintains, all his rights and, like any other citizen, no special privileges. The ensuing situation in years to come would provide the answer to the question not through laboratory or social experimentation but through observation. . . . Legally speaking, such practices as enforced birth control and artificial insemination would interfere with individual freedom, which the author so rightly stresses. . . .

Ingle suggests further that people be paid for barrenness. I suggest instead that the various subsidies of fecundity be halted. . . . This plan would permit the same funds to be channelled in the proper direction, namely, to help those young people in our society who show promise of becoming outstanding adults. Rather than try to modify our unproductive citizenry through impossible frontal assaults, let us concentrate on the boys and girls of talent and ambition. They, in turn, will lead a morally and intellectually healthier society which will develop the means to sustain our growing population and tne wisdom to limit it with justice and which will, as a byproduct, elevate proportionally its lowest stratum, the poor, whom in some form we shall always have with us.

JOHN P. SMITH

326 Cross Street, Fort Lee, New Jersey

Unjustified Fears

... Only two points appear to link the genetic section of Ingle's argument with the sociopolitical discussion in the second half of the article: first, his doubts concerning interracial marriage, and second, the matter of race representation in jobs in the higher echelons. As regards the first, he seems to assume that in interracial marriages the usual practice of choosing a culturally and intellectually suitable partner will be abandoned. As regards the second, he says that, if there are genetically determined racial differences in drives and abilities, then "equal representation of the Negro at the higher levels of job competence and in government will be deleterious to society; return to the principle of judging the employability of the individual without regard to race would be preferred." It seems quixotic indeed to speak of the "return" of the latter principle, and far too early to complain of excessive application of the former, although it may be observed that the rapid establishment of a responsible Negro elite might very well offset minor losses of efficiency (if any), and is certainly a goal not to be summarily dismissed....

MICHAEL A. DEAKIN

5444 Woodlawn Avenue, Chicago

A Question of Relevance

... Ingle states as a "clearly established" point that "Race and color are not valid criteria for judging the worth of an individual. Whatever criteria are needed to judge the individual and to define his rights and freedoms should be applied without regard to race or color." ... If this is so, then race should have no relevance in decisions in giving jobs, housing, and schooling *regardless* of whether there are innate racial differences or not. Group means do not predict individual abilities.... Rights apply only to individuals, not to races or any other collective....

Even more disturbing are his comments that (i) if there are no racial differences, the demand of Negroes for equal representation in government jobs is just, and (ii) if there are differences, then equal representation will be harmful to society. The first statement implies a clear collectivistic premise—racism in reverse. The demand for equal representation by right rather than by merit (that is, a quota system) is not only an extremely racist position but a contradiction of the idea that people are to be judged as individuals. The second position is equally untenable. Equal representation, provided it is the result of the fact that Negroes holding the jobs have earned them, will not be harmful to society. Any quota system based on criteria other than ability will be harmful to society—whether its root is racism or some other irrational doctrine. . . .

EDWIN A. LOCKE

American Institute for Research,
Washington, D.C.

The Province of Science

. . . If the entire review has any scientific content I think it can be fairly summarized as follows:
1)   There is an imprecisely defined quality called intelligence. This quality is known to depend, at least, on inheritance and cultural background. There is at present no known way to evaluate precisely the direct and interaction effects of these factors on intelligence.
2)   Within the population "American" there is a subpopulation "Negro," which is "culturally deprived."
3)   Given this situation, improved measures and precision are required to determine whether the contribution of inheritance to the quality intelligence in the subpopulation "Negro" is greater than, less than, or equal to the contribution of inheritance to this quality in the population "American."

No basis of evidence is given for excluding the first possibility ("greater than") and applying a one-sided test. This perhaps illustrates the extreme problem of experimenter bias when an effort is made to devise tests in matters where there is a deep emotional and sociological content for the population which includes the experimenter.

The question of whether a difference in "genetic intelligence," even if such a difference could be established, would provide a sound and just basis for sociological or political differentiation is not in the proper province of science [unless it is assumed that] science has demonstrated special competence on moral, social, legal, and political questions.

JOHN T. PORTER II

156 Ocean View, Del Mar, Calif.

After reading the letters written in reply to Dr. Ingle's article, have your previous judgments of its worth been strengthened or changed? If so, or if not, why or why not?

As is usually the case with any reputable journal, Science allowed Dr. Ingle a chance to reply to his critics. He did so as follows:*

Several authors of comments on "Racial differences and the future" objected to the publication of views which disagree with their own. The key points of my essay were: (i) The question as to whether the average differences among the races in test performance, school achievement, and behavior have a genetic

---

* *Science*, 18 December, 1964. Reprinted by permission.

as well as environmental basis is unresolved. (ii) The issue is important and should be studied as a means to understanding the causes of social problems and correcting them. (iii) It is time to propose, debate, and test by pilot studies means of preventing social problems, rather than to depend upon palliative methods.

I emphasized that "race" is not a valid criterion for judging the worth of an individual or for depriving him of constitutional rights but claim that the question of average genetic differences among the "races" is important in the struggle for social and biological values. A second important question is this: Is science to continue as the free pursuit of knowledge, or is it to become subordinate to social and political theories which seek to restrict ideas by value judgments rather than by controlled experiments and by logic?

It is commonly claimed that science has proved that there is no genetic basis for the average differences in test performance and school achievement of whites and Negroes. This is what the public, school children, and college students are taught. However, none of the authors of letters claims that this is proved; some admitted that it is unresolved. Why should this uncertainty not be admitted in schools and to the public?

I acknowledge general criticisms as follows:

1) Jaquith, Montagu, and Fried attack my use of the words "race," "intelligence," and others. This is a useful device in debate. Challenge your opponent to define his terms, and, if he falls into the trap, the argument can be kept away from the real question. If "race" and "intelligence" and relevant words can't be defined, ergo, debate and study must stop. There was no evidence that any of my critics were confused by the words nor did they refrain from using them. I do admit to improper use of the word "equalitarian."

2) A method of attack is the use of sarcasm without coming to grips with the point at issue.

3) A means of diverting debate and inquiry is to claim that a proposition is untestable. If science refrained from study of difficult problems, there would be no research on cancer, mental diseases, or several other great diseases.

4) An excuse for attempts to bury the question is that debate will aid the racists. Racists do not accept the principle that each individual be judged according to aptitudes, drives, and behavioral standards without regard to race and that all individuals be granted their constitutional right. I do. I have seen no evidence that racists have made effective use of anything I have said.

5) Finally, it is claimed that any possible average differences in genetic endowment among the races is of no importance from the standpoint of social action. If men were judged solely on the basis of individuality, it would be possible to ignore average differences in aptitudes and drives among the races. Anyone willing to look and learn can find examples of reverse racism not only in the demands of some Negro leaders but in damaging practices already carried out or proposed. I agree with Endler and Deakin that the Negro is still gravely handicapped by racial prejudice and deplore this fact no less than they, but the existence of one form of racism does not excuse the creation of another.

Endler claims, erroneously, that Negroes have never asked for equal representation in jobs. Jaquith seems to support "legally enforced integration" and to imply that equal representation of Negroes at the higher levels of job competence and in government is supported by constitutional guarantees irrespective of competence. Locke disagrees with my opinion that if there are no racial differences in drive and aptitudes the aim of Negroes for equal representation in government is just.

Montagu says that it is a small thing to ask whites to help Negroes of poor behavior to raise their standards. I agree, but propose positive methods to advance the Negro rather than to tear down, level, and destroy good schools and communities by the random forcing of whites and Negroes together. I have lived in a desegregated community for the past $11\frac{1}{2}$ years. The Hyde Park-Kenwood community has, thanks to my university, achieved significant success, when less determined communities and their schools have rapidly become resegregated. I pause to ask Montagu, is Cherry Hill Road integrated? If so, does it include all classes of Negroes? If not, why not?

Here and there, pressure groups have facilitated the random mixing of races, causing enduring harm to neighborhoods and schools, but have failed to prevent resegregation. Whites and Negroes of good behavioral standards retreat from increased crime, filth, and creation of slums. Herein is infringement of the rights of individuals and groups to private judgments, freedom of association, and to life, liberty, and pursuit of happiness. Drive toward excellence of community life—housing, social intercourse, and schools—is a vital form of self-fulfillment dependent upon individual rights and freedoms. These rights and freedoms are not threatened by Negroes of the same standards, but attempts to force the accommodation of all standards within a single community is a destructive form of repression. As Montagu says, the poor behavioral standards of some Negroes are the direct result of the treatment that Negroes have received from the white man. But neither forced desegregation nor forcing the disadvantaged out of urban renewal projects and then forgetting them is the answer. The disadvantaged of all races need special intensive attention.

Groups which seek to force random desegregation are gaining strength without accompanying gains in knowledge or wisdom. If some foreign power had by evil design been able to reduce the white population of the public schools of our nation's capital to less than 15 percent and to make its streets unsafe, America would not have accepted this affront. But the insidious changes caused by unopposed social pressures have been accepted. The schools of other great cities have moved steadily in the same direction to the detriment of all of their citizens. Here and there intelligent plans are emerging which may facilitate voluntary integration while preserving the quality of community life and improving the quality of education. I have in mind the recommendations of the Hauser and Havighurst reports on the schools of Chicago.

There are many examples of successful integration of schools by pupils of compatible abilities and drives. But when it is random or forced, the disadvantaged Negro child is frequently unable to compete. Either the standards of the school are downgraded, or children are grouped according to abilities and Negroes complain bitterly that they are being segregated within the school. Are these enduring average differences in test performance and school progress, which widen each succeeding year, due solely to environment, or do innate differences play a role? Should Negroes expect equality of opportunity to bring equality in achievement? We should make every effort to find out.

I shall answer some specific comments. Jaquith makes the obvious point that there are a variety of positions on race and intelligence. His claim that not all racists maintain that Negroes are genetically inferior surprises me, for I hadn't heard of such. He states that in the absence of firm evidence to the contrary, there is no justification for assuming that racial groups are differentially equipped in respect to genetic potential. I agree that we shouldn't behave as though assumptions are facts, but add that neither is there justification for claiming that races are genetically equal until supported by firm evidence.

Montagu writes approvingly of conception control for all who, either because of genetic limitations or because of poor cultural heritage, are unable to endow children with a reasonable chance to achieve happiness, self-sufficiency, and good citizenship. Montagu and I could quibble over several points, but our only serious disagreement is on his position that the question of race and intelligence is untestable and unimportant. I am glad to see him acknowledge that heredity plays a role in intelligence. Many social scientists teach that intellect is entirely or almost entirely the product of environment and, hence, are unworried about high birth rates among the incompetent.

Rabin, seemingly unworried about the threat of overpopulation, recoils from the recommendations on conception control as being ruthless and inept. He recommends that we not attend to bad culture and behavior, for they are matters of private judgment.

I thank Paula Giese for documenting my claim that such views as hers are held. Here is an expression of doubt that there is a genetic basis for intelligence and a characterization of the proposal that the problem be studied as a mischievous suggestion. She implies that private enterprise should not have a role in upgrading genetic and cultural heritage. The success of integration in the Hyde Park-Kenwood community was achieved largely by private organizations. The integrated housing of Lake Meadows and Prairie Shores and many others was built by private funds. On the national scene, the NAACP, the Urban League, religious organizations, and so on are private enterprises supported by private funds which have facilitated the advancement of the under-privileged.

Fischer doubts that any scientist has proposed that members of another race have an average innate ability superior to those of his race. I am among the non-Jews who consider it probable that superior intelligence and genius occur more frequently among Jews, until recently a disadvantaged people. Jews are less a "race" than Negroes, but races are not randomly represented in this minority group. Fischer imagines that I propose eugenic measures which would select only for intelligence. There are many other important qualities of physique and intellect. I have never proposed a basis for selection.

Fischer and Deakin disagree with my doubts about encouraging interracial marriage. Many integrationists claim that it is not an issue. It is a real and highly sensitive issue, for interbreeding is being encouraged as a means of resolving racial problems. What is wrong with an interracial marriage between culturally and intellectually compatible Negroes and whites? Too little is known of the biological consequences. The question of race and intelligence is unsettled. Less is known of the inheritance of various drives and behavior traits and their relationship to race. We look in vain for a country which is governed wisely by Negroes. Racial mixing cannot be undone. Let's facilitate Negro advancement by full civil rights and equal opportunity, reward and honor their achievements, prevent human misery of every race, but without accepting the social scientist's assurance that the biological experiment of interbreeding can be done without risk to civilization.

Parton complains of my reference to the unpublished studies of Strodbeck. These careful, extensive, and highly significant studies will be published. Strodbeck has kindly given me detailed reports on completed but unpublished phases of the research. I did not, as Parton claims, accept the conclusions of Shuey, but simply mentioned that Shuey and Anastasi had reviewed much the same subject and had reached widely different conclusions.

Each point made by Snow and Seibert was anticipated in my article. In regard to studies on identical twins, I said that "the same studies also demonstrate the importance of environment." I wrote only of a genetic basis for intelligence and made no claim that intelligence is fixed and have never imagined that there is a unitary ability to learn or reason.

I have a final word on the right of the scientist to dissent against attempts to close systems of knowledge. In science we demand validation of each claim to knowledge by rigorous and critical tests of evidence. Positive claims are not final until there is proof that all alternative propositions are untenable. Science does not abdicate to authority or the tyranny of dogma—nor does it try to shape truth by aims and value judgments.

DWIGHT J. INGLE

Department of Physiology, Univ. of Chicago,
Chicago 31, Illinois

*The political implications of Dr. Ingle's articles were difficult to overlook . . . and they were not overlooked, as the following letter indicates.**

Why Equality?

Dwight Ingle, an authority on physiology, has chosen to deal with a topic that is philosophical and political in nature ("Racial differences and the future," 16 Oct., p. 375). I am forced to question his competence as a social thinker.

---

* *Science*, 1 January, 1965. Reprinted by permission.

His heuristic expression of ideas is fraught with emotion, displays a superficial knowledge of current methodology in the field of social reform, and demonstrates serious misunderstanding of the concept of equality as well as the Negro cause.

To argue that people are literally equal is naïve. Few intelligent people do so. But to argue that equality is meaningful only in the context of legal and moral rights is to drastically oversimplify the concept. Adherence to a philosophy of equality is highly useful. Paradoxical as it may seem, it is a philosophy that sustains individuality and nourishes an open and mobile society. It is this aspect of the belief that Ingle ignores.

Let us take just one example—education at its best. If a teacher chooses to accept the concept of equality when confronted with a child who is performing below average, her course is clear. She must devise methods to raise his performance. To do this it is absolutely necessary to meet the individual needs of that child. If he is capable of being "equal" then she must know precisely what has to be remedied to bring this state about. And since all children are, in fact, different, the remedy for one child must differ from the remedy for another. Because this is a philosophy of optimism, the child is always encouraged to attempt more difficult and more varied tasks in an attempt to discover the real depth and breadth of his capabilities. Unless the door is left open, unless we allow children to prove themselves, given our present knowledge we can never know what the maximum potential is. When children are approached otherwise, as is too often the case, they are awarded opportunity according to a predetermined judgment of the teacher, the guard against teacher error in judgment is eliminated, and the result is a shocking attrition of human resources. Perhaps some day science will devise the infallible intelligence test— a test of individual, not group, capacity. When that day arrives, the task of the educator will no doubt be simpler. But until that day arrives we have no choice but to adhere to the conventional wisdom of the times.

Education is but one example. Another is in the area of selection. Unless we act *as though* people are equal, we are faced with the awful task of deciding who is superior. To cite the recent example of Nazi Germany obviates the need for further comment.

It is somewhat confusing to determine the purpose of Ingle's inquiry. What can be gained by knowing whether or not Negroes are inferior? Certainly a knowledge of average group capacity cannot help the educator who must devote his efforts to individual fulfillment. Certainly this knowledge can not help the employer who must hire individuals, not groups. Inquiry into the nature of human behavior and capacity is valid and potentially useful. But Ingle's inquiry is in terms of race. He does not propose a study of social misfits in general, which by his own admission would cut across class and racial lines. His motive seems to be to refute what he considers to be a major voice in the Negro protest—one that cries out for reward without regard to qualification. I strongly object to Ingle's complete misrepresentation of the Negro cause. The demands of Negroes are quite simple. We want to be treated like anyone else. We want our children to know, to the extent that white children know, that the world is open to them. Although there is wise disagreement about means to this goal, the goal most certainly has not been lost in the debate. We are in the throes of a revolution, and revolutions engender violence, anger, and frustration. Heated and irrational views are bound to be expressed, but they do not seem to confuse those white people who are certain of their moral values and have made peace with their fears. The fact that Ingle chooses to stress the arguments of an unrepresentative group reveals his own fear and lack of moral commitment.

Ingle would prefer to have neither forced integration nor forced segregation. This would be lovely, indeed. However, to long for such a happy solution is to resort to the childish device of wishing away reality. Yes, there are communities which have opened their doors to Negroes, and some employers are beginning to question discriminatory hiring policies—these are signs of progress. But the signs are few. The Negro is still faced most often with a choice between remaining in the ghetto or forcing his way out. Perhaps if the Negro waits another generation or two there will be further progress in white morality—this is a moot point. However, the Negro has no intention of waiting; he has no desire to witness his children struggling against the impossible circumstances that have confronted him.

Ingle is perfectly correct to remind us that integration will not solve the Negro's problems. God knows these problems are highly complex and their solution will require the utmost patience and wisdom. But, although integration will not solve the Negro's problems, his problems cannot be solved without integration. Without integration, the Negro has no hope. Unless he *sees* himself living and working among whites, he can only despair that the white world—the successful world—is a world he cannot share.

Ingle longs to achieve equal rights "with a minimum of conflict," which would indicate that he feels threatened by change and is not willing to sacrifice any measure of comfort for the sake of social justice. Revolutions, by definition, involve conflict. Our variety involves "a minimum of conflict" to the extent that relatively few lives are lost and our political structure remains intact.

Ingle proposes various therapeutic social measures, all of which are old hat. The need for nursery education is fully realized by educators and social leaders, and steps in that direction have already been taken. Youth programs most certainly do exist for underprivileged and "bad" boys and girls. I am not sure why Ingle ignores the multitude of settlement houses and similar organizations as well as massive anti-poverty measures which are being taken by federal, state, and local agencies. He has a special admiration for 4-H clubs, which abound in rural areas. Of course in the rural area where Negroes are numerous, namely in the South, 4-H clubs are segregated. Slum clearance, while highly desirable, has been found to serve mainly as a morale booster. It does not cure social ills.

Ingle's final solution seems to be conception control, not for economic reasons, but to prevent reproduction by those "unqualified for parenthood."

The implications of this proposal are political and moral. Ingle would evidently choose to risk a Brave New World rather than to live with the imperfections inherent in a democracy. I would not. To me, individual freedom is sacred. We do spend billions of dollars on crime, delinquency, and similar social ills. If the only alternative to this is to establish a board of judges to decide who is and who is not fit for parenthood, and thus to sacrifice the very heart of our freedom, then I consider these billions of dollars money well spent.

Science is inextricably bound to philosophy and politics. It is no accident that many nuclear physicists have become moral leaders. In our age, when science is both monstrous and wonderful, it is frightening to see among its ranks men such as Ingle, who lack political insight and philosophical discipline.

ADAM C. POWELL

House of Representatives,
Congress of the United States,
Washington, D.C.

*Once again, Dr. Ingle was allowed to reply.**

I invite interested readers to examine my essay on "Racial differences and the future" for evidence that it was "fraught with emotion" and to examine Powell's letter for its relevance to the questions raised by me about biological problems.

Although the concept of equality is not meaningful in biology, I cherish the ideal of equal rights and opportunity for self-fulfillment which extends beyond the opportunity to make material gains to the achievement of dignity and self-respect. The idea that individualization of education aims for equality in achievement is pure demagoguery. Where have we heard it before? I remember: "Every man a king."

Intelligence ranges from idiot to genius among whites, Negroes, and other "races"; and objective tests, imperfect as they are, are sufficiently good to identify the general level of aptitudes and intelligence in individuals. In most cases, it serves the best interests of the child to teach and train him according to aptitudes, interests, and drives.

I have never characterized an ethnic group as "inferior" or "superior." These terms can be meaningfully applied only to individuals. Although it is proper to refer to a genius as being superior in intelligence and a moron as being inferior in intelligence, these terms also connote human value, something that I do not wish to define in terms of intelligence. We would avoid some trouble and misunderstanding by keeping the words "inferior" and "superior" out of debates about average genetic differences among "races."

Contrariwise, and in apparent disagreement with Powell's concept of equality, I recognize differences in human values; the values of what men make of themselves range from the criminal and law evader to the saint, from the demagogue to the statesman, from the indolent to the worker, from the rake

* *Science,* 1 January, 1965. Reprinted by permission.

to the virtuous, from the lout to the gentleman. Judgment of human worth is necessary in a democracy. Shall America accede to those aggressive minorities who cry, "I am equal, give to me according to my wants?" Powell accepts the idea of revolution with conflict aimed at the forcing of integration. He does not admit that the behavior of the average Negro is a critical barrier to integration. He is not willing to guide integration according to individuality but asks that all participate as "equals." I hope for voluntary integration linked with an attack upon the reasons that it is resisted. Racial bias is one. Although larger numbers of Negroes are good neighbors, schoolmates, and employees, many are not. One cause of undesirable behavior is the cultural heritage of the average Negro. If average genetic differences are an important basis of Negro problems, we should have this information to use in guiding Negro advancement.

Powell does not grasp the meaning of my proposal that we aim to prevent the transfer of substandard culture by intensive attention to the child from birth or, better still, beginning with adequate prenatal care. The social measures presently practiced are palliative and feeble. This is one area in which we can learn something from the Soviet Union—without emulating their political aims.

Powell is among those opposed to conception control, even among individuals unqualified for parenthood. (Some of the readers who are not biologists equate conception control with sterilization. The term "birth control" is commonly used, although none of the procedures has anything to do with the process of birth.) Many of the biologically and culturally disadvantaged mate only for pleasure and not for reproduction but lack knowledge of how to control conception. Those imperfections which the Congressman says we should keep in our society are the biological bases of human misery.

Although I hope for the evolution of knowledge and wisdom that will make possible a program of eugenics, I have not imagined that science and society are ready to undertake more than simple educational and advisory programs.

The knowledge of mind and body which we should seek and the methods of preventing human misery which we should debate and test by pilot studies could serve the advancement of all races and especially Negroes. We will not move ahead by saying, "Don't look, don't look, this issue is closed." It is my opinion that if America is guided by Congressman Powell, the role of government in education and social reform will impede rather than facilitate progress, and the Negro ghetto will continue to exist until some of the Negro leaders who are great and wise seek knowledge and truth as the way to freedom.

DWIGHT J. INGLE

Department of Physiology, Univ. of Chicago,
Chicago 37, Illinois

In this controversy, you have read the opinions of persons from the fields of physiology, animal behavior, anthropology, psychology, and, finally, politics. All have become deeply involved. Intelligent laymen have also put forth their opinions. In case you may be inclined to discuss the points raised in this interchange as being a one-shot affair, the following items should serve to dispel this illusion. The first comes from the Stanford University News Service, and is dated April 7, 1967.

A widespread follow-up study of "disadvantaged children" adopted from "improvident backgrounds," aimed at determining whether heredity or environment is more important in their future behavior, has been proposed by Stanford Prof. William Shockley, a Nobel laureate in physics.

"I intend my actions in raising these questions to have the effect of a visitor to a sick friend who strongly urges a diagnosis that seeks to expose all significant ailments," he told a meeting of the National Academy of Sciences here Wednesday (April 26).

"If study shows that ghetto birthrates are actually lowering Negro intelligence, objectively facing this fact might lead to finding ways to prevent a form of genetic enslavement that could provoke extremes of racism," he said. "I feel that no one should be more concerned with this possibility than Negro intellectuals.

"This study could be based on data from adoption programs and should be planned to answer a key question that lies at the very core of the war against poverty:

"Specifically, can improved environment remedy the obviously enormous social disadvantages afflicting the illegitimate 25 per cent of Negro babies? Or will genetic inheritance produce such a low 'social capacity index' that most will perform at frustratingly low social levels?

"If existing adoption programs are so administered that relevant data cannot be obtained," Dr. Shockley added, "I feel confident that eminently humanitarian, appropriately subsidized, voluntary experimental programs could be designed to produce the needed facts.

"These facts could so reduce the present environment-heredity uncertainty about the causes of poverty that they could easily pay off a thousand times their cost in increased effectiveness of programs costing tens to hundreds of billions contemplated over the next 20 years.

"I feel that the government or the large foundations should try to carry out such programs rather than looking away from them as they seem to me to do at present."

In addition, the Stanford engineering professor urged the setting up of a "national study group to do research on what research has already been done. The facts on which definitive conclusions can be based may already be available," he said.

"My call today is for vigorous attempts to establish fact, not for any form of social action.

---

* Reprinted by permission.

"Human quality problems, with their environment-heredity uncertainties and their racial aspects, are on the front pages almost daily," Prof. Shockley continued.

"When someone says: 'Research in this area cannot conceivably be of value to mankind,' it expresses to me an undemocratic contempt for public wisdom that is quite in keeping with totalitarian regimes and wholly out of harmony with the free speech and free press principles of our constitution.

"The lesson to be learned from Nazi history is the value of free speech, not that eugenics is intolerable. A form of eugenics has been in effect in Denmark for 30 years, but I have found no one in this country who has studied it really seriously."

Prof. Shockley said he approached this new (for him) field of research on a "try simplest cases" basis after he discovered that between World War I and today, whites have advanced the equivalent of four I. Q. points relative to Negroes on the armed forces mental tests.

"The facts of behavior genetics and what other information I could obtain led me to conclude that a drop of four I. Q. points in average Negro intelligence could easily have occurred in the course of two generations since World War I as a result of higher birth rates of disadvantaged, improvident people," he said. "My inquiries to eminent anthropologists convinced me further that objective studies were not in progress and were even being discouraged.

"My analysis has led me to devise what I call a 'social capacity index' defined statistically for individuals in a population. It appears to correlate a very diverse assemblage of behaviors ranging from an eminence such as listing in the International Who's Who through study for a law degree, illegitimacy rates, arrests for crime, narcotics addiction, and on down to commission of murder.

"In respect to whites, U.S. Negroes perform at about 10 times lower frequency in achieving eminent accomplishment and at nearly 10 times higher frequency in anti-social behavior such as narcotics addiction and murder. In contrast, Chinese and Japanese Americans are almost equally offse whites, at least in regard to science achievement and arrests.

"The universal pattern employed is essentially the one that has been found for scores on I.Q. tests. This observation suggests that I.Q. test results may actually be a deeper measure, at least on a statistical basis, of a distribution of some more fundamental social capacity.

"I do not conclude that my studies prove that a genetic offset actually exists," Prof. Shockley warned. "The conclusions are consistent with a model that assumes that a genetic offset, equivalent to about 18 I.Q. points, is the principal cause.

"Until acceptable facts are established, the truth or falsity of the predominantly genetic view is obscured by the environment-heredity uncertainty," he concluded. "I deplore this uncertainty."

"If environment is the main cause, the present uncertainty will inhibit our overcoming unreasonable prejudice. If genetics is the main cause, the uncertainty will cloud public discussions and search for solutions.

"Furthermore, vast expenditures in our well-intentioned war on poverty may accomplish not a solution, but instead create a larger problem—a situation comparable to providing economic aid to underdeveloped countries while disregarding the population explosion."

Another report on this topic appears in the April 27, 1967, issue of the New Scientist,* a British journal comparable, perhaps, to Scientific American:

Are Some Races Cleverer?

Few other socio-scientific subjects are as controversial as the question of whether there exists a genetic basis for a difference in intelligence between races. There are politicans who would welcome scientific proof that Negro children are less intelligent than their white contemporaries, and there are others who would rather believe that the Earth is flat than that such a difference exists.

Two papers given yesterday at the 104th meeting of the National Academy of Sciences tried to take a dispassionate look at the whole wretched battlefield. One fact that cannot be disputed—both speakers used it as the springboard for their discussions—is that the intelligence of Negro children in the United States —as measured by standard IQ tests—is at least an order of magnitude lower than the national average. Wallace A. Kennedy of Florida State University quoted a study of 1800 representative Negro school children from the Southeast which listed achievement and intelligence in 1960 and followed up a portion of the sample in 1965. It found a 20-point deficit in intelligence of Negro children, a deficit that stayed constant throughout their school career, and remained stubbornly the same even after a sharply improved education.

Before these facts are seized upon in triumph by racialists (and to illustrate the difficulties which anthropologists are up against), other observations have shown that, at birth, and during the first two years of life, tests of mental intelligence (of necessity, measured by other means) do not show this deficit in Negro intelligence. Several studies have even demonstrated a slight superiority of Negro infants at birth.

These, then, are the observations; what of the conclusions? Here there are two main schools of thought. The first says that, since so many racial characteristics—such as height, weight, skin pigment, facial characteristics, and even susceptibility to disease—are genetically determined, why should not intelligence be included?

---

* Reprinted by permission.

The second view maintains that intelligence is dependent upon cultural, not genetic factors. One study by a subscriber to this school demonstrated that for certain young, deprived children, intelligence was modifiable by as simple a thing as nursery school attendance. Under this banner, of course, comes the theory that diet—an integral part of environment—plays an important part in the determination of intelligence (see "Notes on the News," Vol. 32, p. 498).

The great stumbling block to the resolution of the conflict is inherent within the subject: scientists are plain scared of it. Unless a researcher in the United States, for instance, clearly announces before he even starts that he subscribes to the second school of thought it is blasphemy to investigate differences. "And, although no one in the twentieth century is struck dumb for blasphemy", as Dr. Kennedy pointed out this week, "his research funds can be struck, and the effect is the same."

Both Kennedy, and the second speaker, Professor William Shockley of Stanford, outlined their ideas for the methods to be used in solving the heredity versus environment conundrum. The facts, particularly as they relate to the cause of the paralysing poverty among Negroes in the United States, "could easily pay off a thousand times [the cost of gathering them] in increased effectiveness of programmes costing tens to hundreds of billions of dollars contemplated over the next 20 years", according to Shockley. Kennedy's "definitive study" would select experimental and control subjects from among illegitimate children who are wards of court. The mothers should have been well cared for in maternity homes. The child would then be placed in a middle-class Negro home, so that he would receive all the advantages and opportunities of a middle-class environment. They would be followed up during their schooldays, on the objective criteria of intelligence, achievement and their position in class. Shockley's proposal is similar.

And, finally, the National Academy of Sciences makes an official statement:

RACIAL STUDIES: ACADEMY STATES POSITION ON CALL FOR NEW RESEARCH

William Shockley of Stanford, who won the Nobel prize for work on transistors, has lately been arguing for an expansion of research to evaluate the relative effects of heredity and environment on human intelligence and performance. Implicit in his proposals is at least the speculation that inferior genetic inheritance, rather than inferior environment, accounts for the relatively poor performance of some Negroes in various competitive situations. Specifically, Shockley has been calling for a study of "disadvantaged children" who have been adopted from "improvident backgrounds." As he put it in a talk last spring to the National Academy of Sciences, the object of the study would be

---

* *Science,* 17 November, 1967. Reprinted by permission.

to answer the question, "can improved environment remedy the obviously enormous social disadvantages afflicting the illegitimate 25 percent of Negro babies? Or will genetic inheritance produce such a low 'social capacity index' that most will perform at frustratingly low social levels?"

Shockley's vigorous advocacy has been a matter of some discomfort to the Academy, which finds itself situated between its traditional belief in free inquiry and its realization that the formulation of heredity versus environment adds up to a loaded question that might be destructively exploited by racists if the Academy even ratified it as the right question. At the Academy's fall meeting on 23 October, in Ann Arbor, President Frederick Seitz presented the NAS's Council's response to Shockley's proposals, though, in fact, the statement made no direct reference to Shockley himself. The Academy statement, which was prepared with the assistance of several geneticists (James F. Crow, Wisconsin; James V. Neel, Michigan; and Curt Stern, University of California, Berkeley) follows:

The Academy has been urged to take strong measures to reduce the present uncertainty about the relative importance of heredity and environment as causes of human social problems and as causes of racial differences in behavioral traits. It is asked to promote actively the seeking of answers to such questions as: To what extent are urban slums the result of poor heredity? Is the genetic quality of the human population being seriously eroded by economic and medical advances that have dramatically decreased the death rate, and by differential birth rates in various social, economic, and educational groups? Are genetic factors responsible for a significant part of racial differences in educational and economic achievements? Could a eugenic program materially reduce our major social problems? By concentrating on environmental approaches, is society neglecting promising genetic possibilities?

The question has been raised as to whether research in these areas is being carried out as vigorously and intelligently as it should be.

Do anthropologists and geneticists have an environmentalist bias that discourages research into the hereditary bases of individual and racial differences in intelligence and ability to adapt to our society? Is this research being seriously impeded by investigators' fears that the results might be unfavorable to some ethnic minorities?

How urgent is it that such questions be answered?

We certainly need to know more about human genetics; as to the desirability of further research there can be no serious question. Researchers in experimental and human genetics have brought deep insights concerning ourselves and our past. The detailed understanding of the molecular basis of heredity is one of the intellectual triumphs of the twentieth century. New genetic knowledge is already bringing practical benefits in the understanding, prevention, and treatment of genetic diseases. We can expect continued rapid progress in this area.

With complex traits like intelligence the generalities are understood, but the specifics are not. There is general agreement that both hereditary and environmental factors are influential; but there are strong disagreements as to their relative magnitudes—which is another way of saying that the evidence is not conclusive. Furthermore, it is not obvious that really substantial increases in this knowledge will come soon, even if the amount of research were greatly increased. The problem of disentangling hereditary and environmental factors for complex intellectual and emotional traits where many genes may participate, where measurements are often not reproducible, where it is not certain what is being measured, and where subtle environmental factors are involved is extremely difficult. It is unrealistic to expect much progress unless new methods appear.

Even greater difficulties are encountered in any attempt to assess the relative role of heredity and environment in determining racial differences in intellectual and emotional traits. Despite the great number of tests that have been performed on Negro and white populations, it is still not clear whether any differences found are primarily genetic or environmental. For example, there is no scientific basis for a statement that there are or that there are not substantial hereditary differences in intelligence between Negro and white populations. In the absence of some now-unforeseen way of equalizing all aspects of the environment, answers to this question can hardly be more than reasonable guesses. Such guesses can easily be biased, consciously or unconsciously, by political and social views.

It is indeed possible that some studies have not been carried out for fear that the results might not be acceptable to some groups. Many researchers prefer to work in noncontroversial areas where public feelings are not involved and where they can work undisturbed. There is, however, a more valid reason that might keep scientists from working in such areas as the separation of hereditary and environmental contributions to complex human behavioral traits and to racial differences in these traits. This is the conviction that none of the current methods can produce unambiguous results. To shy away from seeking the truth is one thing; to refrain from collecting still more data that would be of uncertain meaning but would invite misuse is another.

Yet, it is not proper to say that we know nothing about the inheritance of complex traits, or that the consequences of a genetic program are not at all predictable. Animal experiments have shown that almost any trait can be changed by selection. The immensely successful history of animal and plant breeding, for a long time based on no more complicated principle than that "like begets like," shows this. A selection program to increase human intelligence (or whatever is measured by various kinds of "intelligence" tests) would almost certainly be successful in some measure. The same is probably true for other behavioral traits. The *rate* of increase would be somewhat unpredictable, but there is little doubt that there would be progress.

On the other hand, it is contrary to all evidence that social problems such as poverty, slums, school dropouts, and crime are *entirely* genetic. There is surely a substantial and perhaps overriding environmental and social component. Therefore, society need not wait for future heredity-environment research in order to attempt environmental improvements, nor will it do so. We can be sure that no amount of genetic research will demonstrate the futility of *all* attempts at environmental improvements. It should be emphasized that the existence of even a strong hereditary component in any condition, individual or social, does not imply that the condition cannot be cured or ameliorated.

There are two aspects of eugenics that, although not entirely different, are sufficiently distinct to be considered separately. They are:

1) *The reduction of the incidence of known inherited diseases.* This involves the discovery by medical, chemical, or cytological techniques of persons with a high risk of having children with gross abnormalities, or with severe physical or mental disease. A great deal of human misery, both of parents and of children, can be prevented through genetic counseling. The decisions can be made by the individuals involved; social decisions are ordinarily not needed.

2) *Attempts to alter the population genetically for intellectual and emotional traits that vary continuously, or to reverse possibly undesirable effects of differential fertility.* To bring about any substantial change in the next generation would require a large change in reproductive patterns. To do this by education, by persuasion, by economic incentives, or by stronger measures would require social decisions that are not lightly made.

It is clear that for many important and complex traits the population could be changed by either genetic or environmental means. They are not mutually exclusive; more likely they are mutually reinforcing.

Heredity-environment uncertainty is not the main reason for avoiding drastic selection measures. The major impediment to eugenic action is not genetic ignorance but rather Society's uncertainty about its aims and about the acceptability of the means for attaining them. Even if it were known beyond doubt that the heritability of social maladjustment is very high, would we choose to remedy the situation by eugenic means?

For one thing, our society still severely restricts even the voluntary individual application of some available techniques. Birth control is only gradually becoming legally accepted, especially among the unmarried, long after it has become widely practiced among well-to-do and educated people. Therapeutic abortion is very safe when done under proper medical conditions, but is forced underground or to other countries, with the consequence that it is available safely only to the privileged. Artificial insemination, although widely practiced, is in such a questionable legal position that no accurate records, even of its frequency, are available. Any program of genetic improvement, even if entirely voluntary, would be seriously impeded by inability to make full use of techniques now known.

Moreover, regardless of the acceptability of the methods and regardless of the success of research in disentangling the role of heredity and environment in complex social traits, society is far from ready to interfere to any significant extent with the reproductive preferences of this generation in order to change the gene pool of the next. On the other hand environmental measures have wide and immediate social acceptability.

Genetic changes are measured in generations. Whatever genetic deterioration is occurring as a result of decreased natural selection or by differential birth rates is slow relative to many environmental changes, particularly those associated with technological innovations. Likewise, genetic improvement by any eugenic program that is likely to be accepted in the near future by our society would also be slow.

For these reasons, we question the *social* urgency of a greatly enhanced program to measure the heritability of complex intellectual and emotional factors. This is not to say that such work should not be done. But we would not, for example, urge that work in other parts of genetics be reduced in order to supply trained personnel to study this area more intensively.

Likewise, we question the social urgency of a crash program to measure genetic differences in intellectual and emotional traits between racial groups. In the first place, if the traits are at all complex, the results of such research are almost certain to be inconclusive. In the second place, it is not clear that major social decisions depend on such information; we would hope that persons would be considered as individuals and not as members of groups.

On the other hand, no promising new approach to answering these questions should be discouraged. While existing methods offer little hope for unambiguous answers, there is always the possibility that new insights will come from an unexpected direction. The history of scientific discovery suggests that the best strategy would be the support of basic research from which such insights may arise.

Here, then, is an area which lies well within the purview of scientific investigation. Yet, by virtue of its obvious sociological and political implications, the objectivity necessary to make an investigation truly a scientific one becomes difficult to obtain.

It is not very taxing to think of other such areas. Biological investigations on insecticides, industrial pollutants, smoking, etc., indicate their hazards to man's health and his surrounding ecosystem. Legislative action brought about by these discoveries often runs into strong opposition from groups who would be affected by such legislation. Similarly, the biological dangers of overpopulation suggest the need for the wider dissemination of birth control information. Such legislation, however, runs into stiff opposition from certain politically powerful religious groups.

It would be a mistake, however, to automatically assume that all such organizations which oppose Federal legislation leading out of scientific research findings are necessarily acting for purely selfish motives, or are simply wrong. For example, legislation, probably arising out of the thalidomide scare,* made it difficult (and occasionally impossible) for research investigators to obtain certain other drugs for legitimate research purposes.

---

* Thalidomide caused severe congenital malformations to children whose mothers had taken the drug during early pregnancy.

A second mistake would be to assume that for matters with both scientific and political ramifications, there is such a thing as a "scientific viewpoint." The brilliant chemist Linus Pauling wholeheartedly supported the nuclear test ban treaty (designed, among other things, to lessen the amount of fallout radiation to which the biosphere is exposed). The equally brilliant physicist Edward Teller opposed the treaty. The unanimity of agreement among biologists concerning the validity of the concept of evolution by natural selection fast disappears when these same biologists consider political issues—a fact which is hardly surprising.